THE U.S. TECHNOLOGY SKILLS GAP

Founded in 1807, John Wiley & Sons is the oldest independent publishing company in the United States. With offices in North America, Europe, Asia, and Australia, Wiley is globally committed to developing and marketing print and electronic products and services for our customers' professional and personal knowledge and understanding.

The Wiley CIO series provides information, tools, and insights to IT executives and managers. The products in this series cover a wide range of topics that supply strategic and implementation guidance on the latest technology trends, leadership, and emerging best practices.

Titles in the Wiley CIO series include:

The Agile Architecture Revolution: How Cloud Computing, REST-Based SOA, and Mobile Computing Are Changing Enterprise IT by Jason Bloomberg

Big Data, Big Analytics: Emerging Business Intelligence and Analytic Trends for Today's Businesses by Michele Chambers, Ambiga Dhiraj, and Michael Minelli

The Chief Information Officer's Body of Knowledge: People, Process, and Technology by Dean Lane

CIO Best Practices: Enabling Strategic Value with Information Technology by Joe Stenzel, Randy Betancourt, Gary Cokins, Alyssa Farrell, Bill Flemming, Michael H. Hugos, Jonathan Hujsak, and Karl D. Schubert

The CIO Playbook: Strategies and Best Practices for IT Leaders to Deliver Value by Nicholas R. Colisto

Enterprise IT Strategy, + Website: An Executive Guide for Generating Optimal ROI from Critical IT Investments by Gregory J. Fell

Enterprise Performance Management Done Right: An Operating System for Your Organization by Ron Dimon

Executive's Guide to Virtual Worlds: How Avatars Are Transforming Your Business and Your Brand by Lonnie Benson

Innovating for Growth and Value: How CIOs Lead Continuous Transformation in the Modern Enterprise by Hunter Muller

IT Leadership Manual: Roadmap to Becoming a Trusted Business Partner by Alan R. Guibord

Managing Electronic Records: Methods, Best Practices, and Technologies by Robert F. Smallwood

On Top of the Cloud: How CIOs Leverage New Technologies to Drive Change and Build Value Across the Enterprise by Hunter Muller

Straight to the Top: CIO Leadership in a Mobile, Social, and Cloud-Based World (Second Edition) by Gregory S. Smith

Strategic IT: Best Practices for Managers and Executives by Arthur M. Langer

Strategic IT Management: Transforming Business in Turbulent Times by Robert J. Benson

Transforming IT Culture: How to Use Social Intelligence, Human Factors, and Collaboration to Create an IT Department That Outperforms by Frank Wander

Unleashing the Power of IT: Bringing People, Business, and Technology Together by Dan Roberts

The U.S. Technology Skills Gap: What Every Technology Executive Must Know to Save America's Future by Gary J. Beach

THE U.S. TECHNOLOGY SKILLS GAP

WHAT EVERY TECHNOLOGY EXECUTIVE MUST KNOW TO SAVE AMERICA'S FUTURE

Gary J. Beach

Cover image: © Mick Wiggins/Alamy
Cover design: Michael Rutkowski

Published by John Wiley & Sons, Inc., Hoboken, New Jersey.
Published simultaneously in Canada.

For general information on our other products and services or for technical support, please
contact our Customer Care Department within the United States at (800) 762-2974, outside
the United States at (317) 572-3993 or fax (317) 572-4002.

Wiley publishes in a variety of print and electronic formats and by print-on-demand. Some
material included with standard print versions of this book may not be included in e-books or in
print-on-demand. If this book refers to media such as a CD or DVD that is not included in the
version you purchased, you may download this material at http://booksupport.wiley.com. For
more information about Wiley products, visit www.wiley.com.

Library of Congress Cataloging-in-Publication Data:
Beach, Gary J., 1950–
   The U.S. technology skills gap : what every technology executive must know to save
America's future / Gary J. Beach.
      pages cm. — (CIO series)
   Includes bibliographical references and index.
   ISBN 978-1-118-47799-1 (cloth); ISBN 978-1-118-66044-7 (ebk.); ISBN 978-1-118-66047-8
(ebk.); ISBN 978-1-118-68070-4 (o-book); ISBN 978-1-118-79232-2 (custom)
1. High technology industries—United States.   2. Labor supply—United States.   3. Skilled
labor—United States.   4. Vocational qualifications—United States.   5. Information
technology—United States.   6. Science—Study and teaching—United States.   I. Title.
   HC110.H53.B43 2013
   338'.0640973—dc23

                                                                                    2013007117

To the 49,266,000 schoolchildren in America's public schools and their futures.

CONTENTS

PART TWO—And the Hits Just Keep on Coming   83

CHAPTER 9 The Skills Gap Warnings Begin   87

CHAPTER 10 The Skills Gap Emerges   111

CHAPTER **11** The Skills Gap Widens **121**

CIOs SPEAK

Most books have forewords authored by one individual who often explains his, or her, passion for the topic covered by the book. For this book I decided to go a different route and invited 16 chief information officers to share their opinions about the importance of the skills gap challenge facing our nation. Their statements follow.

• • •

"Unless we build a stronger curriculum in science, technology, and math and raise our expectations for K–12 education, we will foster a generation of tech-savvy *users* with few skills to *build* or innovate technology. The results will be detrimental to our country and our potential ability to compete in the global digital economy."

Adriana Karaboutis, Vice President and Global CIO, Dell Inc.

"Success in IT requires a mastery of the fundamentals underpinned by strong 'C' skills: critical thinking, collaboration, and communication. Our best people apply critical thinking to determine how emerging technologies can be harnessed to deliver value for clients, ever mindful of changing marketplace and business requirements."

Frank B. Modruson, CIO, Accenture

"America has a rich tradition of making things. The increasing technical sophistication of the world, combined with historically low numbers of science, technology, engineering, and math (STEM) graduates, at best fails to honor that history. And at worst it threatens to severely limit America's future."

Ralph Loura, CIO, Clorox Company

"In the past few years I have hired many deeply technical people. The vast majority of résumés for my most technical jobs come from graduates of colleges in India and China. It is clear to me that we are not preparing American students with the skills that high-tech employers deem necessary."

John Halamka, CIO, Beth Israel Deaconess Medical Center, and Professor, Harvard Medical School

"When I talk to high school and college students, I find that the connection of the skills learned in math and science to the skills used in work and life is missing. Educators need to make this connection. How does a lab in science relate to work and life? How does calculus relate? The lack of these connections is a serious gap in our education system."

Nancy Newkirk, CIO, International Data Group

"Information technology plays a pervasive and critical role in driving business capabilities and enabling corporate strategies. In order for American industry to sustain its renowned capacity to innovate, it must have a workforce equipped to develop and apply future generations of advanced information technologies."

James Nanton, Senior VP and CIO, Hanesbrands Inc.

"The American educational system has lost touch with the reality of providing people with the practical skills and competencies required for young professionals to add meaningful value to our corporations. America needs to rethink how we prepare young people to have meaningful careers that are both financially and intellectually rewarding."

Larry Bonfante, CIO, U.S. Tennis Association

"One of the most difficult roles I have as a chief information officer is finding and recruiting talent. In a growing business, with average turnover rates, I run at a constant talent deficit because I cannot find people with the skills I need to fill the job openings I have. If the American education system cannot produce a workforce with the appropriate skills, then these jobs will be filled by global providers. The need to focus on creating career-ready individuals is not an educational imperative. It is an economic imperative."

Gary King, Executive VP and CIO, Chico's Inc.

"The K–12 years are critical foundational years that 'plant the seed' for a desire to learn, to teach vital study and research habits, to develop skill sets, and to discover areas of interest and proclivity. These are pivotal years that work to shape the whole person. The K–12 educational phase is also the ideal period to generate interest in and a desire and passion for technology. Sadly, more and more of our underserved demographic groups are participating as consumers of technology rather than as developers or innovators of such."

Gina C. Tomlinson, CTO, City and County of San Francisco

"I became astutely aware that America had a problem communicating and getting children interested in technology based on an experience I had with my middle school–age daughter, who told me one day, 'Dad, I am terrible in technology.' The first thing I told her, partly kidding, was not to say that in public too loudly, because that would not look good for Dad, since his job is heading a technology group! But it illustrated a problem our country has: most children are not being exposed to the possibilities of technology; to how the field could be interesting, challenging, and a great job opportunity for them; and to the fact that they should not have any fears about being able to utilize technology in many ways, since they already use it far more than they comprehend."

Michael Gabriel, Executive VP and CIO, Home Box Office

"The historical position of the United States as a global technology innovator has brought us prosperity and growth. These effects will dry up quickly, however, if our country does not produce a steady supply of thinking leaders who are able to compete in the global technology marketplace. As our world shifts more and more from atoms to bits as the currency of economic growth, America will be left behind if we are not able to compete as global innovators. As a result, we will soon find ourselves handing our global economic leadership over to a new set of leaders and, along with it, our ability to determine our own future and control our own destiny. The United States must make profound, wholesale changes to our education system in a way that emphasizes science, technology, engineering, and math (STEM) and encourages and motivates students to excel in these critical areas. If we fail to do so, we will lose our global competitiveness."

Steve Mills, CIO, Rackspace Hosting Inc.

"'Survival of the fittest' has shaped the evolution of our species for hundreds, thousands, even millions of years. In the twenty-first-century business context, the fittest are those with the ability to think critically, solve problems, innovate, and collaborate effectively with one another. If we fail to equip our children with these skills through significant enhancements to our education systems, how will they ever survive?"

Bill Schlough, Senior VP and CIO, San Francisco Giants

"I highly encourage and support the preservation of a technologically strong America through education. An influx of human talent into the science, technology, engineering, and math fields is necessary to accelerate the innovation that will ultimately change companies, people, and society for the better."

Thaddeus Arroyo, CIO, AT&T

"The shortage of qualified resources in the technology and engineering sector has weakened the job market and the talent pool of the American workforce. As a CIO, I have a much tougher time finding qualified candidates today compared to 20 years ago. This shortage of qualified staff is forcing businesses to outsource more work to developing markets."

Atti Riazi, CIO, New York City Housing Authority

"The United States has a storied history of invention and innovation that fueled its twentieth-century journey to become a global economic and military power. Working at a federal government research and development center for 35 years, I have become more sensitive to the importance of technical innovation, particularly information technology, to the security of our country. But today we find ourselves losing ground to competing countries in science, technology, engineering, and math education, and with it our technology leadership. These are trends we must reverse. It is truly a matter of national security."

Gerald R. Johnson, former CIO, Pacific Northwest National Laboratory

"Not that long ago, America's system of education was considered the world's incubator of innovation, but sadly we have lost our dominance in this area. Fortunately for America, we can correct our course, but it will require cooperation from parents, faculty, industry, government, and students. If we fail to do so, the American Dream will regrettably remain a Dream Deferred."

Tony Coba, Senior VP and CIO, Miami Heat

PREFACE

These educational gaps impose on the United States the economic equivalent of a permanent national recession.

—MCKINSEY & COMPANY[1]

In a country that spends $583 billion each year on public education, the taxpayer deserves a better return on investment.[2] For nearly two decades, America's fourth-, eighth-, and twelfth-grade students have performed poorly in math and science compared to their peers in other countries. Over a slightly longer time frame, as our country's education policy shifted to accountability under President George H. W. Bush, the results in domestic math and science assessment tests have been worse: SAT scores for math have stagnated since the 1980s, and verbal scores are now the worst on record for the SAT.

I have no "street cred" as an educator, although I did teach theology to high school freshmen in my first job out of college in 1972. But 30 years of conversations with information technology (IT) executives does afford me a small soapbox to step up on and broadcast loud and clear an escalating point of pain they shared with me: America's schools are not producing individuals with the strong quantitative and communicative skills necessary to compete in the twenty-first-century global economy.

The skills landscape has changed significantly in America over the past 173 years. In 1840, 79 percent of the American labor force worked in the agricultural and manufacturing sectors.[3] Only 21 percent were employed in service jobs. By 2010, the composition did a near-complete reversal, with 88 percent of American jobs in services and 12 percent in agriculture and manufacturing.[4] Critics of the American school system claim that what, and how, we teach schoolchildren is largely based on the 1840 percentages. And a 2012 McKinsey and Company report flatly states that "a skills shortage is a leading reason for entry level vacancies that cause significant problems [for American firms] in terms of cost, quality, time, . . . or worse."[5] The Computing Technology Industry Association reported in 2012 that a whopping 93 percent of employers indicated that there is an overall skills gap.[6]

Reading and literacy was the key subject for American education in the 1950s. Rudolph Flesch, in his 1955 best-selling book, *Why Johnny Can't Read: And What You Can Do about It*, touched a raw nerve with the American public. Fifty-eight years later, the key skills for American schoolchildren are math and science proficiency. But no one has yet authored *Why Johnny and Mary*

Can't Do Math and Science. Although many reports have addressed this subject, American students, parents, teachers, politicians, and teacher union leaders have not worked effectively together to build a collaborative solution.

I do. And as I reviewed decades of results of key math and science assessment tests, I kept asking myself this question: How did America get to the point where we invest more money than any other country on the planet in public education only to get mediocre returns? Surely it could not be a purposeful national strategy. Are American kids in the twenty-first century genetically less gifted than students in Finland, Singapore, Japan, and South Korea? Surely not. So something else must be the reason. I set out to find out what it was.

The twenty-first-century American education system is a vast enterprise, built on factory-floor management techniques of the late nineteenth century with the goal of mass-producing as efficiently as possible the 49,266,000 students who come through the doors of nearly 100,000 public schools each year.[7] As an article in *Public Interest* magazine noted, the American education system is focused on "mediocracy rather than meritocracy."[8]

Here's a vivid example of what that means. In 1993, as publisher of the newspaper *Computerworld*, I was master of ceremonies at its annual Search for New Heroes program with our partner, the Smithsonian Institution. One of our award winners that year was Seymour Papert, the founder of the well-known Massachusetts Institute of Technology Media Lab. Papert was being honored for his work in education, and as he joined me at the podium to accept the award, he looked out into the audience of 1,500 black-tie and evening-gown technology executives and asked the producer to raise the lights so he could see the crowd. He then asked, "By a show of hands, how many of you in this audience at your business promote workers based on age?"

Standing next to Papert, I looked out into the crowd and saw a sea of puzzled faces. Not one person had raised his or her hand. But Papert wasn't finished. He commented, "In two weeks [the Computerworld-Smithsonian event was always held on the first Monday in June], three million third graders will be asked to stand up and march out the door, with their teachers bidding them a fond farewell and 'good luck in fourth grade.'" Papert's point: Business doesn't promote by age, and neither should educators.

There are three important reasons that America must improve the math and science skills of its students. The first is economic. In 2009, McKinsey and Company researched the results of the international math and science assessment tests, created a proprietary measurement tool, and reported that if America's students matched the proficiency of Finland's students in math and science, the U.S. gross domestic product would be 16 percent larger each year.[9]

The second reason is employment. With the world connected by millions of miles of fiber cable that allow work to be done anywhere on the planet,

Americans are no longer competing with the person in the next cubicle for their next promotion or job. Their competition for work in the twenty-first century will be tens of thousands of miles away. The U.S. unemployment rate will remain historically high as businesses look to hire the best skills for the best price.

Right now, America is losing that battle, according to a January 2012 report published by the Harvard Business School. It found that 58 percent of senior business leaders in America believe that the nation's K–12 education system is currently worse than that of other countries, and nearly 80 percent of these leaders said that the American education system is continuing to fall behind that of other countries.[10]

The third reason to improve math and science skills is just starting to surface: the national security of our country. Throughout the history of humanity, the arenas of war have been ground, sea, air, and space. Now, in the twenty-first century, add cyberspace to the list. To protect our country's financial, utility, and defense infrastructures, we must produce more, rather than fewer, students with strong skills in math and science.

In 2007 I began the journey that resulted in this book. As I started my project, I created a folder on my computer called "A Country Left Behind" (a purposeful reference to the high-profile No Child Left Behind Act) and filled it with thousands of documents on tests, industry warnings, and general observations of what needed to be done. It was one of those files that when you open it, it gives you a headache because you often can't find what you are looking for, there is so much in it.

As I struggled to develop the outline of my story, I had an idea. What if I tell the story of why American kids are performing poorly in math and science through a time line that roughly parallels my life? I was convinced that American students were not genetically less gifted than their international peers, so maybe if I went back to around 1950, the year I was born, I could tell the story—and, more important, connect the dots that explain our nation's poor performance—through events that have occurred over roughly the past 70 years.

I reviewed my "A Country Left Behind" file and organized the most relevant documents in decade folders starting in 1940, and that's when my story unfolded in my head and on my keyboard!

I learned that as far back as 1909, Americans had an aversion to studying math and science. I learned that the math skills of the men who wanted to enlist in the army in 1941 were so abysmal that the army had to construct a test to minimize the effects of the American education system. I learned that President Franklin D. Roosevelt tried to get the nation broadly focused on the importance of science in American life, but he failed. I learned that America looked the other way after World War II and embraced Nazi scientists and engineers and put them in charge of our country's missile program, which culminated in Neil Armstrong and Buzz Aldrin landing on the moon in July 1969. I learned why at the height of the Cold War in 1958, the U.S. Department of Education, desperate for any guidance in improving math and science education in our country, sent two delegations to Moscow to study how the Soviets taught those subjects!

The story continued as I discovered how an American engineer single-handedly altered the global economy and why possibly the current ills of American math and science education just might be blamed on teachers who taught the baby boomers. I learned why the work rules embedded in teacher union collective bargaining contracts that are more than 100 pages long have a stranglehold on the American education system. I learned about the theory of an obscure Japanese physicist called "Yuasa's Phenomenon," which predicts that America's best days may be behind it. I learned that from 1963, the first year that American baby boomers took the Scholastic Aptitude Test (SAT), through 1976, both math and verbal SAT scores declined each year. Yet nothing was done. And I learned that the creation of the U.S. Department of Education had nothing to do with education but everything to do with a political quid pro quo offered to the National Education Association, the big teachers' union.

My research connected me to scores of reports from government, industry, and academia with ominous sounding titles like *A Nation at Risk*, *Gathering Storms*, and *Quiet Crisis*. Reports began to be released in 1983, and they delivered the same dire message: something is systemically wrong with America's education system. These reports were incredibly well written. They convinced me that the American education system had missed the sea change from a manufacturing-based global economy to a service-based economy and that improving public education in the future has nothing to do with dollars spent.

Before I started the project, I knew about the SAT. But I was not aware of two other international math and science tests: Trends in International Mathematics and Science Study and the Programme for International Student Assessment. I knew about "teaching to the test," but I had no idea that the "test" was called the National Assessment of Educational Progress, self-proclaimed by the U.S. Department of Education as the "nation's report card." I gathered pertinent information on all of these tests, and I have presented them in chronological order so the reader can see how American students have been developing since 1964. It is not a pretty picture.

Finally, I learned about an incredible array of great work that IT workers, their companies, and nonprofits are doing to complement and supplement the teaching of math and science in American schools. I am proud to share with you examples of the best programs I found, and I hope to inspire you to create one at your company or to join an existing program.

During the early phase of my research, I read a famous speech that Sir Winston Churchill delivered to the House of Commons in November 1936.

ACKNOWLEDGMENTS

Writing a book is not an easy task, nor is it done in isolation. During the six years it took me to research and write *The U.S. Technology Skills Gap*, there were thousands of people I spoke with individually about the topic or exchanged ideas with through my columns in *CIO* magazine and on CIO.com. Although I cannot possibly mention them all by name, their ideas and recommendations have made their way into the pages of the book.

Some people, however, do deserve individual acknowledgment for their contributions.

For her guidance, recommendations, and patience during the editing phase of the project, I want to thank Stacey Rivera, my development editor at John Wiley & Sons. Thanks also to Tim Burgard, my acquisitions editor at John Wiley & Sons for his support and encouragement, which made this book a reality. And thanks to Kimberly Monroe-Hill, senior production editor at Wiley, and Judith Antonelli, my copyeditor, who molded my words into the book you are about to read.

I must thank my friend Geoff Smith, the former deputy chief information officer at Procter & Gamble, for inviting me to a workforce development conference in the middle of the winter in 2007; that's where I had the seminal idea to write this book. And even though I have never met them, Titus Galama and James Hosek, two economists from the Rand Corporation, deserve my thanks for introducing me to the fascinating story of Mitsutomo Yuasa, which is a central theme throughout the book.[1]

A book takes on a life of its own. Especially during the research phase, when manila file folders are scattered about one's home, the dining room table becomes a large rectangular filing cabinet, and family members, including pets, become a focus group of sorts, forced to listen to ideas about the book.

My wife, Catherine, was an incredible partner during all phases of the book. Her patience was remarkable, and only once did she ask me to clear the dining room table: on the day before Thanksgiving! I must thank my son, Scott, who has an engineering degree and works as a manager for Accenture, for serving as a personal role model of what a successful career in business and technology is all about. And special acknowledgment must go to Noelle, my daughter, who was my North Star as I wrote the book. Her daily input was extremely valuable.

Finally, I must offer a very special thanks to those who teach. Lee Iacocca, the former automobile executive, once said, "In a completely rational society, the best of us would be teachers and the rest of us would have to settle for something else."

Here's to hoping that this "completely rational society" arrives soon in America.

Note

1. Titus Galama and James Hosek, *U.S. Competitiveness in Science and Technology* (Santa Monica, CA: Rand Corporation, 2008).

PART ONE

How Did
We Get Here?

I n 2000, Malcolm Gladwell wrote a best seller entitled *The Tipping Point: How Little Things Can Make a Big Difference*. The subtitle serves well to introduce Part One of *The U.S. Technology Skills Gap*.

As I mentioned in the preface, I was curious why American schoolchildren in the twenty-first century were performing so poorly compared to schoolchildren from other countries. It certainly wasn't genetics, so some other factor, or combination of factors, must be the reason. But what were those factors? Were they related, or were they random? And once they were identified, would it be possible to identify a solution? Or had America's education system long ago passed the point of no return?

I will let you make that decision once you have read my analysis in this section of the book.

CHAPTER 1

1941: The Subject We Love to Hate

For every 1,000 students in fifth grade, 600 are lost to education before the end of high school.

—National Science Foundation[1]

That twenty-first-century American students perform poorly on international math and science assessment tests compared to the rest of the world should not be a surprise. For more than a century, Americans have shown little interest in, or mastery of, these subjects—especially math.

Though justifiably proud of our country's long list of accomplished inventors and entrepreneurs like Thomas Edison, Henry Ford, and Steve Jobs, Americans are clueless about our nation's legacy of math and science heroes. Americans continue to be awarded the vast majority of Nobel Prizes in the twenty-first century for physics, chemistry, medicine, and economics—fields that demand a solid understanding of science and math—yet the average American has no idea who these modern-day science and math heroes are.

Want proof? Ask a colleague this question: "Who is the most famous American scientist?" Most people will pause for a moment and then say, "Albert Einstein." Technically, they would be correct. For the last 15 years of his life, from 1940 to 1955, Einstein was a naturalized American citizen, after fleeing the threat of Hitler and the Nazi Party in Germany in the mid-1930s. But Einstein, a German by birth, was educated in Switzerland; he never spent one minute sitting in an American classroom studying math or science. The U.S. education system cannot claim Albert Einstein as its poster boy.

Want to have more fun? Ask someone else this question: "Who is the most famous American mathematician?" No one has a clue—and a friend of mine actually replied, "Russell Crowe." (Crowe, of course, is an actor who played the part of the economics Nobel Prize winner John Nash in the 2001 film *A Beautiful Mind*.)

Math? Not for Me!

Today's American students hail from a long lineage of young Americans who seem to run the other way when math is involved. To support that claim, David Klein, in a book about the history of K–12 math education, compiled a list showing the percentage of U.S. high school students enrolled in algebra, geometry, and trigonometry at various times from 1909 to 1934. Table 1.1 shows what he found.[2]

What makes the numbers in Klein's list so notable is that the significant ongoing decrease in the percentage of American students studying math happened just as an amazing array of new technologies was dramatically affecting life in America—inventions and product innovations like Henry Ford's Model T automobile (1908), talking motion pictures (1910), the radio circuit (1918), liquid-fuel rockets (1926), the jet engine (1930), and Polaroid photography (1931).

Given the surge in interest in science and math that occurred during the dot-com boom in the late twentieth century, I would have expected more, rather than fewer, students opting to study math from 1909–1934. But that didn't happen. Some people point to the Great Depression as the primary reason that fewer young Americans studied math in that period. Their rationale is that schoolchildren not only opted out of studying math but also opted out of attending school, in order to go to work to help keep their families financially afloat.

But that conjecture is wrong. Look at Table 1.1 again. By far, the majority of slippage occurred from 1909 through 1928, one year *before* the stock market crash of 1929 that ushered in the Great Depression.

Americans just don't like math.

Table 1.1   Slip Sliding Away

Percentage of High School Students Enrolled in Math			
School Year	Algebra	Geometry	Trigonometry
1909–1910	56.9%	30.9%	1.9%
1914–1915	48.8%	26.5%	1.5%
1921–1922	40.2%	22.7%	1.5%
1927–1928	35.2%	19.8%	1.3%
1933–1934	30.4%	17.1%	1.3%
Percent change 1909–1934	(–46.6%)	(–44.7%)	(–31.6%)

"Minimize the Effect of Schooling"

Regardless of whether Americans liked math, in 1934 a development in Europe was occurring that would jolt America awake to its shortcomings in math and science education.

On August 2, 1934, Adolf Hitler became the fuehrer of Germany and accelerated the massive manufacturing rearmament of that country—an effort highly dependent on a German workforce with strong science, technology, engineering, and math skills. In the next six years, German workers designed and built the world's first jet fighter (the Messerschmitt), massive naval ships (like the battleship Bismarck), advanced rocketry, and highly sophisticated tanks.

As Germany was rearming, America was still reeling from the effects of the Great Depression. Few dollars were available for investment in military technology. In fact, the U.S. Army had more mules (56,000) than tanks (464) in 1940.[3] But the tanks, aircraft, and communications equipment the army did own in 1940 were far more complex than the comparable equipment used at the end of World War I.

The Japanese attack on Pearl Harbor ushered in America's participation in World War II, and as 10 million men and women signed up to be inducted into the armed services, the U.S. War Department realized it had a big personnel problem.

The War Department needed intelligent recruits to train as officers and to operate modern weapons correctly. Klein described the personnel problem this way: "In the 1940s it became something of a public scandal that army recruits knew so little math that the army itself had to provide training in arithmetic needed for basic bookkeeping and gunnery."[4]

To solve this problem, the army created an intelligence aptitude test called the Army General Classification Test. When the test was introduced in 1941, the army said it was needed to ascertain the verbal comprehension and quantitative reasoning skills of the recruits and to "minimize the effect of schooling" in the classrooms of America.[5]

The word *minimize* is not a typo.

Young Adults with IQs of Eight-Year-Olds

The Army General Classification Test grouped the verbal and quantitative test scores of the recruits into five categories, ranging from fast learners to slow learners. Between March 1941 and December 1942, 4,569,451 men and women took the test. Table 1.2 shows the results of the initial examination.[6]

Table 1.2   Fit to Serve?

Category	Number	Percent
1 (fast learners)	275,206	6%
2 [not named]	1,169,591	25.6%
3 (average learners)	1,381,460	30.2%
IV4 (below average learners)	1,174,543	25.7%
5 (slow learners)	568,615	12.4%

Rod Powers, writing about the Armed Services Vocational Aptitude Battery (ASVAB), a test now used by the military to determine the eligibility of a recruit, claimed that the 1941–1942 test takers who were classified in categories 4 and 5—38 percent of those tested—had cognitive verbal and quantitative reasoning skills of *eight-year-old third graders*.[7]

That, too, is not a typo!

The Fall Continues

After the war, the long-term decline in American students studying algebra, geometry, and trigonometry shown in Table 1.1 mostly continued (see Table 1.3).

The final 46-year enrollment tally from 1909 was as follows: algebra, down 56 percent; geometry, down 63 percent; and trigonometry, up 37 percent.

Table 1.3   Math Interest Mostly Continues to Wane

Percentage of High School Students Enrolled in Math			
Year	Algebra	Geometry	Trigonometry
1934	30.4%	17.1%	1.3%
1949	26.8%	12.8%	2.0%
1953	24.6%	11.6%	1.7%
1955	24.8%	11.4%	2.6%

President Roosevelt Understands Science

One person who understood the importance of science and technology to American life was President Franklin D. Roosevelt. On November 17, 1944, Roosevelt instructed Vannevar Bush—a Massachusetts engineer and close confidant who served as the president's science advisor and as a key manager of the Manhattan Project, which built the first atomic bomb—to write a special report to "do what can be done to make known to the world as soon as possible the contributions which have been made during our war effort to scientific knowledge, the new frontiers of the mind that are before us, and if they are pioneered with the same vision, boldness and drive with which we have waged this war we can create a fuller and more fruitful employment and a fuller and more fruitful life."[8]

Roosevelt did not live to read Bush's report, entitled *Science: The Endless Frontier*, which was delivered to President Harry Truman in July 1945. (Yes, *Star Trek* fans, the title of Bush's report was paraphrased by producer Gene Roddenberry for the famous voice-over opening of every episode: "space, the final frontier.")

The introduction to the report is worth reading:

Some of us know the vital role which radar has played in bringing the United Nations to victory over Nazi Germany and in driving the Japanese steadily back from their island bastions. Again it was painstaking scientific research over many years that made radar possible.

What we often forget are the millions of pay envelopes on a peacetime Saturday night which are filled because new products and new industries have provided jobs for countless Americans. Science made that possible, too.

In 1939 millions of people were employed in industries which did not even exist at the close of the last war—radio, air conditioning, rayon and other synthetic fibers, and plastics are examples of the products of these industries. But these things do not mark the end of progress—they are but the beginning if we make full use of our scientific resources. New manufacturing industries can be started and many older industries greatly strengthened and expanded if we continue to study nature's laws and apply new knowledge to practical purposes.

Advances in science when put to practical use mean more jobs, higher wages, shorter hours, more abundant crops, more leisure for recreation, for study, for learning how to live without the deadening drudgery which has been the burden of the common man for ages past. Advances in science will also bring higher standards of living, will lead to the prevention or cure of diseases, will promote conservation of our limited national resources, and will assure means of defense against aggression. But to achieve these objectives—to secure a high level of employment, to maintain a position of world leadership—the flow of new scientific knowledge must be continuous and substantial.

Our population has increased from 75 million to 130 million between 1900 and 1940. In some countries comparable increases have been accompanied by famine. In this country the increase has been accompanied by more abundant food supply, better living, more leisure, longer life, and better health. This is, largely, the product of three factors—the free play of initiative of a vigorous people under democracy, the heritage of great national wealth, and the advance of science and its application.

Science, by itself, provides no panacea for individual, social, and economic ills. It can be effective in the national welfare only as a member of a team, whether the conditions be peace or war. But without scientific progress, no amount of achievement in other directions can insure our health, prosperity, and security as a nation in a modern world.[9]

President Truman embraced Bush's report and proposed four key science initiatives, aimed mostly at the high-end scientific community: (1) increase overall science funding, (2) emphasize the importance of basic research, (3) create the National Science Foundation, and (4) grant additional funds to colleges and universities for research.

Truman felt strongly that "continuous research by our best scientists is the key to American scientific leadership and national security. Now and in the years ahead, we need, more than anything else, the honest and uncompromising common sense of science. When more people have learned the ways of thought of the scientist, we shall have better reason to expect lasting peace and a fuller life for all."[10]

These were all worthy initiatives. But they were aimed at benefiting America's "best scientists," not the masses of poorly educated Americans.

An Opportunity Lost

Science: The Endless Frontier hinted that Bush and Truman were at least aware of the abysmal state of the overall American education system at the close of World War II:

The country may be proud of the fact that 95 percent of boys and girls of the fifth-grade age are enrolled in school, but the drop in enrollment after the fifth grade is less satisfying. For every 1,000 students in the fifth grade, 600 are lost to education before the end of high school, and all but 72 have ceased formal education before completion of college. While we are concerned primarily with methods of selecting and educating high school graduates at the college and higher levels, we cannot be complacent about the loss of potential talent which is inherent in the present situation.

Students drop out of school, college, and graduate school, or do not get that far, for a variety of reasons: they cannot afford to go on; schools and colleges providing courses equal to their capacity are not available locally; business and industry recruit many of the most promising before they have finished the training of which they are capable.[11]

Those are remarkable words coming from America's leaders only 68 years ago.

Why were America's education goals set low enough to allow for public acclamation of the fact that 95 percent of fifth graders were still enrolled in school? How could the president of the United States not be appalled that 60 percent of American students dropped out of formal schooling before the end of high school? Why didn't someone realize that our country had a huge education problem?

Neither public education nor mathematics and science education was a national priority at the end of World War II in the United States. President Truman was, however, keenly aware of how far behind other nations, specifically Germany, the American education system was at the time. This topic is addressed in Chapter 2.

Americans Still Hate Math and Science

On October 15, 2009, a *New York Times* article interviewed five experts on why American students were faring poorly in math assessment tests in the United States. The comments posted by readers on the web site version of the article are more insightful than the article itself. Here are two readers' opinions of why Americans continue to do poorly in math:

Has it ever occurred to anyone that students don't do well in math because they hate it? Or is that too obvious? People don't tend to do well in a subject that they hate. Plus everyone else (teachers, parents, television characters) show an equal disdain for math. As long as our culture puts *zero* emphasis on math, students are going to see learning math as a chore, not a door to their future.

There is no silver bullet so stuff your "Japanese schools do this" and this theory does that. Face the facts: not everyone can do math![12]

Glen Whitney, a mathematician who worked as an algorithm manager on Wall Street before founding the Museum of Mathematics, the country's first math museum, sums up well America's arm's-length relationship with math. In an April 2012 Reuters interview, he says, "America has a cultural problem with math. It's the subject, more than any other, that we as a country love to hate. We don't see it [math] as dynamic. It's rote and boring and done by dead Greek guys a thousand years ago."[13]

But it's also done, and exceptionally well, by students in scores of countries around the world in the twenty-first century—a fact that now directly threatens the future of America's economic vitality, global workforce preparedness, and national security.

And a fact that forced the president of the United States in 1945 to reluctantly import Nazi scientists and engineers into America and assign them to a mission that ended with the greatest scientific feat in the history of humanity.

Notes

1. Vannevar Bush, *Science: The Endless Frontier* (Washington, DC: U.S. Government Printing Office, July 25, 1945), 23.
2. David Klein, *A Brief History of American K–12 Mathematics Education in the 20th Century* (Charlotte, NC: Information Age, 2003), 175.
3. "U.S. Horses and Mules during WWII," Olive Drab, www.olive-drab.com.
4. Klein, *A Brief History*, 175.
5. A. W. Tamminen, *A Comparison of the Army General Classification Test* (Minneapolis: University of Minnesota Veterans Administration Guidance Center, 1950), 646.
6. Ulysses Lee, *United States Army in World War II* (Washington, DC: Center of Military History of the United States Army, 1966), 243.
7. Rod Powers, *ASVAB for Dummies* (Hoboken, NJ: John Wiley & Sons, 2010), 8.
8. Franklin D. Roosevelt to Vannevar Bush, November 17, 1944, Special Collections, Valley Library, Oregon State University, Corvallis, http://osulibrary.oregonstate.edu/specialcollections.
9. Bush, *Science: The Endless Frontier*, 7.
10. Harry S. Truman, "Presidential Address before the American Association for the Advancement of Science," September 13, 1948, Harry S. Truman Library and Museum, www.trumanlibrary.org.
11. Bush *Science: The Endless Frontier*, 23.
12. "How to Improve National Math Scores," *New York Times*, October 15, 2009, http://roomfordebate.blogs.nytimes.com/2009/10/15/how-to-improve-national-math-scores.
13. Stephanie Simon, "Numbers Game: America's Struggle to Make Math Fun," Reuters, April 11, 2012, www.reuters.com/article/2012/04/11/us-usa-education-math-idUSBRE83A18220120411.

CHAPTER 2

1945: Operation Paperclip

America's First War for Tech Talent

Men who were classified as "ardent Nazis" were chosen—just weeks after Hitler's defeat—to become respectable citizens.

—THOMAS BOWER, THE PAPERCLIP CONSPIRACY[1]

Ask any information technology (IT) worker this question: When did the practice of outsourcing begin? Most will mention the massive remedial software code projects outsourced to India in the years leading up to 2000. The more astute will go back further and cite the 1989 landmark deal in which the Eastman Kodak Company outsourced its entire corporate IT infrastructure to the IBM Company and the Digital Equipment Corporation.

Both answers, however, are wrong.

The first outsourced tech project in American technology history was named after a simple office instrument, the paper clip. It began in September 1945 and involved hundreds of imported German scientists, many of whom were Nazi Party members and even SS officers.

Nazis Hailed as "Outstanding" Scientists

As captured German scientists began to arrive in America in 1945, the U.S. War Department's Bureau of Public Relations, sensitive to a brewing public relations disaster, issued a press release in October about "outstanding German scientists" that read, "The Secretary of War has approved a project whereby certain German scientists are being brought to this country to ensure we take full advantage of exploiting German progress in science and technology."[2]

These "certain" and "outstanding" Germans were part of Operation Paperclip, a military and intelligence project to recruit German scientists to work for the United States after the war. The name derives from the fact that new biographies were written for these individuals to cover up their Nazi pasts, and these expunged versions were paper-clipped to the other records in their personnel files.

Operation Paperclip was considered necessary because at the end of World War II, America was decades behind the rocket science and other engineering efforts of Germany. (The operation was also intended to prevent the scientists from being recruited by the Soviet Union.) As the project unfolded, it did not seem to matter to American leaders that many of the key individuals were alleged senior members of the Nazi Party and the Schutzstaffel, or SS, the elite force headed by Heinrich Himmler that was responsible for horrific crimes against humanity.

The U.S. War Department said that the forced importation of "outstanding" German scientists was needed for national security reasons. But the *real* reason was this: In the first half of the twentieth century, the American education system, as described in Chapter 1, was doing a horrendous job teaching math and science to American children. Fewer than one in 10 young Americans earned a college degree at the time, and 60 percent of high school students dropped out. America's pipeline for future scientists, engineers, and mathematicians was barren.

Germany's Rocket Man

Operation Paperclip's central figure was Wernher von Braun, the world's most talented rocket scientist at the end of World War II. Von Braun was born in Wirsitz, Germany, in 1912 and moved to Berlin, where as a 17-year-old he became fascinated with space travel after reading Herman Oberth's *Die Rakete zu den Planetenraumen* (By Rocket into Interplanetary Space). He attended the Technical University of Berlin and became a member of Verein fur Raumschiffarhrt, the Spaceflight Society, and there told Auguste Piccard, then the world's most famous high-altitude balloonist, that he planned on traveling to the moon one day. Von Braun did not know it at the time, but the most notable accomplishment of his future career would indeed involve interplanetary travel—to the moon.

Germany was focused on rearmament in the early 1930s, and Germans with strong science and engineering skills like von Braun were quickly drafted into the military. As a 21-year-old, just as Hitler's National Socialist German Workers Party came to power, von Braun joined the German army, the Wehrmacht, and was assigned to the rocket center in Peenemünde, a remote village on the Baltic Sea.

At Peenemünde, von Braun drew the attention of Walter Dornberger, the Wehrmacht general responsible for developing Germany's rocket program. Dornberger was impressed with von Braun's scientific and engineering skills and arranged for von Braun to receive state-allocated research grants to fund his rocket-building projects. Von Braun soon learned there was a catch, however. Dornberger informed von Braun that to continue to receive the research grant funds he would have to become a member of the Nazi Party. Von Braun decided that continued work on rocket development was more important than politics and so joined the Nazi Party:

> I was officially demanded to join the National Socialist Party. At this time (1937), I was already technical director of the Army Rocket Center at Peenemünde . . . my refusal to join the party would have meant I would have to abandon the work of my life. Therefore I decided to join, though my membership in the party did not involve any political activities.[3]

The Nazis Get to von Braun

Rocket technology was the overriding passion for this 25-year-old. If he had to be a member of the Nazi Party, a political group that was creating widespread political and military chaos throughout Europe at the time, so be it. But the pressure was just starting to be put on von Braun. Three years later, in the spring of 1940, as Hitler and the Nazi Party waged war across Europe, senior party officials again visited von Braun and presented him with a far more foreboding decision: join the SS, or your rocket research project will immediately be shut down.

Here is how von Braun described it: "One SS Colonel Muller looked me up in my office at Peenemünde and told me that Reichsführer-SS Heinrich Himmler had sent him with the order to urge me to join the SS. I called immediately on my military superior, who informed me that if I wanted to continue our mutual work, I had no alternative but to join."[4]

Smithsonian historian Michael J. Neufeld, who in 2007 wrote a book on the life of von Braun, claims that the previous comment by von Braun cannot be thoroughly verified: "As with von Braun's party membership, we have no truly independent account of what happened, but his story is plausible."[5]

Heinrich Himmler was a key member of Hitler's inner circle and was the commander of the system of concentration camps in which 6 million Jews and approximately 5 million other people were murdered. As a member of the SS from 1940–1944—he eventually reached the rank

Figure 2.1   Wernher von Braun with Nazi Officials at Nordhausen
Source: www.cynical-c.com

of captain—von Braun led Germany's efforts to design and build military rocket weapons, first with a team of scientists at the Peenemünde research site and later at a new underground rocket construction facility in Nordhausen, located in the Harz Mountain region of Germany (see Figure 2.1).

As von Braun's rocket research efforts reached the critical production stage at Peenemünde, a problem had arisen: the German army needed a protected underground rocket construction facility to be quickly built so that Allied bombing raids would not destroy it. Thus the facility at Nordhausen was built, using prisoners from the Dora concentration camp nearby as slave labor. An estimated 20,000 concentration camp inmates were worked to death constructing the tunnels. It has been reported that a prominently placed sign at the entrance of Dora read: YOU COME IN THROUGH THIS GATE AND YOU LEAVE THROUGH THAT CHIMNEY (i.e., of the crematorium).[6]

Time Magazine Paints a Dim Picture of von Braun

A 2002 *Time* magazine article directly connected Wernher von Braun to the use of slave labor in Dora and claimed that he was part of the brutality:

> For reasons best known to von Braun, who held the rank of colonel in the dreaded Nazi SS, the prisoners were ordered to turn their backs whenever he came into view. Those caught stealing glances at him were h[anged].[7]

Other reports place von Braun at the actual concentration camp, which he visited often to discuss with its officials how to get the maximum amount of work out of the prisoners. Tom Gehrels, a University of Arizona professor, makes a claim in the *Time* magazine article that further implicates von Braun: "Dora inmates remember von Braun arriving in the morning with an unidentified woman, having to step between bodies of dead prisoners and under others still hanging."[8]

By the summer of 1944, von Braun and his development team of about 500 scientists had moved to the deployment stage of the V-2 rocket, the world's first long-range ballistic missile. On September 8, 1944, three months after D-Day, Hitler made a decision to launch the first of 3,000 V-2 rockets at targets in Paris, London, Antwerp, and The Hague, killing thousands of innocent people in those cities and creating waves of terror throughout Europe.[9] These rockets were designed and built by Wernher von Braun.

America's Best Rocket: The Bazooka

At this point in World War II, just how far advanced were German rocket scientists compared to their American counterparts? Pictures tell the story. Figure 2.2 shows the V-2 rocket and the U.S. Army's best rocket, the bazooka.

The V-2 could fly at a speed of 3,500 miles per hour at an altitude of 55 miles and an operational range of 200 miles.[10] The tank-destroying weapon, the bazooka, was a shoulder-mounted rocket with an operational range of 300 to 650 yards. The Germans were so far ahead of the rest of the world that the basic technology of the V-2 rocket was not equaled for more than a decade after the conclusion of World War II.

A point of irony, and credit, must be added. America did have the *original* rocket man. Robert H. Goddard, a physicist, created and built the world's first liquid-fuel rocket in 1926, which could fly at an altitude of nearly 2 miles and

Figure 2.2   The V-2 and the Bazooka: Not Even Comparable
Source for v2 rocket: www.wehrmacht-history.com/heer/missles/v-2-rockets-launched-against-england.htm
Source for bazooka: www.scoopweb.com/bazooka

at a speed of 550 miles per hour.[11] Goddard also invented the bazooka during World War I for the U.S. Army.

But Goddard's lifelong ambition, to launch a rocket that could reach the moon, was publicly criticized in venues like the editorial page of the *New York Times*, which said that Goddard didn't understand Newton's Third Law of Motion: "Professor Goddard, with his 'chair' in Clark College and the countenancing of the Smithsonian Institution, does not know the relation of action to reaction. He seems to lack the knowledge ladled out daily in high schools."[12] Goddard was upset with the public ridicule and retreated from public view, which greatly limited the influence of his work.

Back in Germany, von Braun's Faustian bargain with the Nazis was falling apart. Himmler ordered the Gestapo to arrest von Braun for publicly criticizing the V-2 rocket efforts, but another prominent Nazi, Albert Speer, the minister of armaments and war production, interceded directly with Hitler for him, and von Braun was not put in prison.

In 1945, however, the situation worsened for von Braun. As the Third Reich was collapsing, Hitler, concerned that von Braun and his scientists could be captured by the Allies or the Soviets, changed his mind and ordered the Gestapo to execute von Braun and his team of scientists and engineers. Von Braun learned of his fate, and as Germany's remaining military forces fought back the Soviets, who were rapidly approaching Germany from the east, he picked 500 of the top German rocket scientists and hid in an abandoned old mineshaft in the Nordhausen facility with their highly confidential blueprints for building rockets. The Allies captured them in May 1945.

Shipped to America

Through the summer of 1945, the Allies interrogated thousands of these captured German scientists. Low-value prisoners of war were allowed to go home. The most knowledgeable were shipped to America in Operation Paperclip and were forced to sign temporary work contracts with the U.S. Army Ballistic Missile Agency. These "temporary" work contracts lasted the rest of their lives.

Albert Einstein, who personally knew many of the captured German scientists and was keenly aware of their Nazi allegiances, pleaded with President Truman not to allow them to immigrate to America. But Truman didn't, or politically couldn't, listen to Einstein's plea. America desperately needed the scientific and technological knowledge that von Braun and his team of German scientists brought to America.

The egregious acts of von Braun's tainted Nazi past would be ignored by America. Von Braun and his team were allowed to go about their work for the U.S. Army Ballistic Missile Agency, designing and building the most advanced rockets in the world.

America Had Space Technology before the Soviets

Here's a little-known historical footnote: so advanced was von Braun's work that on September 20, 1956, more than a year *before* the Soviet orbital launch of *Sputnik*, von Braun and his team of German scientists (most of whom were now naturalized American citizens) successfully launched a four-stage Jupiter-C rocket from Cape Canaveral in Florida that reached a speed of 13,000 miles per hour and a height of 862 miles and could have easily launched an object into orbit around Earth.[13]

President Dwight Eisenhower was concerned that if the U.S. Army were the first American group to launch an object into orbit, it would be overtly challenging to the Soviets, so he ordered von Braun to put sand in the rocket's fourth stage. In essence, von Braun gave the United States the opportunity to be the first country in the history of the world to launch an object into space, but we chose to pass.

Wernher von Braun would have to wait another 13 years for fame and the fulfillment of his lifelong interplanetary dream, when the massive Saturn V rocket that he and his team built took Neil Armstrong, Buzz Aldrin, and Michael Collins to the moon in July 1969.

Von Braun's work was so important to America's global image, however, that now, instead of posing with Nazi leaders, he stood next to, and briefed, the president of the United States (see Figure 2.3).

Figure 2.3   Von Braun with President Kennedy
Source: NASA

Germany Developed the Atomic Bomb First

Operation Paperclip, America's first outsourced IT project, is not something American IT workers can look back on with pride.

Operation Paperclip was necessary because the American education system—in which millions of students opted out of studying subjects like math for the first 45 years of the twentieth century, which tolerated a 60 percent high school dropout rate, and which was proud of the fact that 95 percent of fifth graders were still in school—failed the country, plain and simple.

As World War II ended, America was decades behind most German technology—including the atomic bomb. According to a 2005 review of the book *Hitler's Bomb*, as early as 1944, "Hitler had the atomic bomb first, which he called the 'wonder bomb,' and tested the weapon, which had roughly the power of a terrorist 'dirty bomb,' on concentration camp workers."[14]

Fortunately for America, Einstein was working on the U.S. atomic weapon in the Manhattan Project. And the Germans could not work out a way to deploy their bomb aerially. Had German engineers been able to do that, imagine how different the world would be today!

Operation Paperclip put America in the unenviable position of abandoning its moral conscience. It placed our country's future national security in the hands of several hundred captured Nazi scientists.

America was able to fulfill President John F. Kennedy's promise to send a man to the moon by the end of the 1960s because of the strength of the German science and math education system—not America's.

But 6,800 miles west of Washington, DC, a young, talented American mathematician, educated from kindergarten through graduate school in America's education system, was about to forever alter the balance of power in the global economy.

Notes

1. Thomas Bower, *The Paperclip Conspiracy: The Hunt for the Nazi Scientists* (Boston: Little, Brown, 1987), 2.
2. U.S. War Department, "Outstanding German Scientists Being Brought to the U.S.," press release, October 1, 1945, National Archives, http://history.msfc.nasa.gov/german/war_department.pdf.
3. Wernher von Braun, U.S. Army affidavit of membership in the Nazi Party, June 18, 1947, National Archives, RG330, 709A4398.
4. Ibid.
5. Michael J. Neufeld, *Von Braun: Dreamer of Space, Engineer of War* (New York: Knopf, 2007), 121.
6. Holocaust Education and Archive Research Team, "Joseph Bau's Journey through the Past," Holocaust Research Project, www.holocaustresearchproject.org/othercamps/plaszow/bauplaszow.html.
7. Leon Jaroff, "The Rocket Man's Dark Side," *Time*, March 26, 2002.
8. Ibid.
9. T. Dungan, "Timeline for V-2 Attack 1944–1945," V2Rocket, www.v2rocket.com/start/deployment/timeline.html.
10. Owen Edwards, "Wernher vonBraun V-2 Rocket," July 2011, Smithsonian.com.
11. Clark University Robert H. Goddard Library, Additional Resources, Archives and Special Collections, www.clarku.edu/research/goddard.
12. "Aeronautics Now," http://astronauticsnow.com/history/goddard/index.html, retrieved February 4, 2013.
13. "Covering the Past, Present and Future of Cape Canaveral," Spaceline, 2012, http://spaceline.org/rocketsum/jupiter-c.html.
14. Ernst Gill, "Hitler Won Atomic Bomb Race but Couldn't Drop It," *Sydney Morning Herald*, March 5, 2005, www.smh.com.au/news/World/Hitler-won-atomic-bomb-race-but-couldnt-drop-it/2005/03/04/1109700677446.html.

CHAPTER 3

1950: Deming Says

Failure of management to plan for the future and to foresee problems has brought about waste of manpower, of materials, and of machine-time, all of which raise the manufacturer's cost and price that the purchaser must pay. The consumer is not always willing to subsidize this waste. The inevitable result is loss of market. Loss of market begets unemployment. Loss of market and unemployment are not ordained. They are not inevitable. They are man-made.

—W. Edwards Deming, *Out of the Crisis* [1]

As World War II ended in Europe, another story was developing in the war in the Pacific. This one involved a native-born American with strong science, technology, engineering, and math skills. So strong, in fact, were his mathematical skills that he single-handedly changed the balance of the global economy.

W. Edwards Deming was born on October 14, 1900, in Sioux City, Iowa, and moved to Wyoming as a young child. His family was frugal, so in his youth he learned a trait that would serve him well throughout his life: never waste anything. After graduating from the University of Wyoming with a degree in electrical engineering, he moved to Colorado and earned a graduate degree in mathematics and physics. He later earned his doctoral degree in mathematics from Yale University. Deming's first job was building telephones for Western Electric, where he learned a lesson that would define his career: the importance of product uniformity in the manufacturing process.

Deming Has an Idea

After his stint at Western Electric, Deming headed to Washington, D.C., where he worked for the U.S. Department of Agriculture before getting a position at the U.S. Bureau of the Census, where he applied his theory of statistical sampling to the collection of 1940 census data. Before 1940, the census in America was conducted door-to-door by nearly 90,000 individuals called enumerators. It was a long, laborious process.

Deming had a better idea. He suggested that the Census Bureau should complement door-to-door census taking with his mathematical sampling techniques. Deming's approach worked brilliantly. With fewer doors to knock on and fewer census takers needed, the 1940 U.S. census was conducted faster and less expensively than ever before.

The War Department noticed. Although many of what journalist Tom Brokaw called America's greatest generation have died, history books and stories passed down through households have told how Americans put conservation and preservation of key materials like metal, aluminum, and rubber at the top of our nation's priority list. In such an environment, census star W. Edwards Deming used his keen understanding of math and statistics to teach military leaders how the government could use what he called total quality management techniques to more efficiently produce war goods.

Total quality management is a manufacturing technique that singularly focuses on iterative, continuous improvement of the quality of products and processes. Deming was ordered by the army to create a special eight-week course for senior military leaders in Washington on how to apply total quality management techniques to the production of war materials. Those leaders, in turn, took the power of Deming's techniques to military bases around the world, and 31,000 people were quickly trained to think and act according to those techniques, which can be broadly summarized in four words: plan, do, check, act.[2]

In 1947, senior leaders of the War Department ordered Deming to join General Douglas MacArthur's Allied powers' team in Japan. His assignment was to conduct an accurate census of the war-ravaged Japanese population in order to assess the proper amount of materials needed to rebuild homes, schools, hospitals, and roads throughout the country. From all the written accounts of his experiences in conducting the Japanese census, Deming became enamored with the Japanese culture and with Japan's science and technology leaders.

The Lecture Series That Changed the Balance of the World Economy

After the Japanese census was completed, Deming returned home. But he didn't stay there for long. The most prestigious scientific organization in Japan, the Japanese Union of Scientists and Engineers, invited him back to Japan for a 12-city lecture series on the principles of total quality management and how those theories could accelerate the reconstruction of the Japanese economy.

Deming accepted and returned in June 1950. On his first 11 stops on the lecture tour, he was well received. The last stop, however, would change forever the balance of the global economy.

For that final lecture, in August 1950, Deming traveled to the Mount Hakone Conference Center, about a two-hour train ride outside Tokyo. What made this lecture important was not its content; it was the same lecture as the previous 11 lectures. What made the twelfth stop special was the *audience*. The previous 11 Deming lectures had reverberated through the halls and corner offices of every major Japanese manufacturing company. Japan's industrial base was still struggling so much that several firms had moved key production facilities to a Japanese town called Usa just so they could affix MADE IN USA tags to their products.[3]

Business leaders in Japan were desperate for guidance, and the audience at the Mount Hakone Conference Center consisted of 75 chief executive officers of the largest and most well-known Japanese companies, including Akio Morita, the cofounder of the Sony Corporation. In his own words, here's the key advice W. Edwards Deming shared with the leaders of Japan's business community:

> As all modern-day manufacturers are striving to make their business prosperous in the long term, the following issues are necessary: 1) better design of products to improve service, 2) higher level of uniform product quality, 3) improvement of product testing in the workplace and in research centers, and 4) greater sales through side [global] markets.[4]

If they adhered to these principles of total quality management, Deming told these business leaders, the following would happen:

- Costs will go down.
- Producers can economize on raw materials.
- Production levels will increase and waste decrease.
- Product quality will become more uniform.
- Producers and consumers will gain the ability to agree on product quality.
- Quality will be improved, so inspections may be reduced.
- Appliances and techniques can be used to a higher degree.[5]

Japanese business and scientific leaders embraced Deming's message. Four months later, in December 1950, the Japanese Union of Scientists and

Engineers created the Deming Prize to honor the Japanese company that best followed Deming's advice. This distinguished prize continues to be awarded annually in Japan.

Japan Embraces, America Ignores

By the end of 1951, the concept of *kaizen*, a Japanese word meaning "improvement" or "change for the better," became commonplace on factory floors throughout Japan. And 19 years later, in 1970, these Japanese manufacturers began taking market share away from American companies in automobiles, office automation, electronics, and steel. This was a result of the free advice given by W. Edwards Deming to Japan's business leaders in the August 1950 lecture.

Why didn't America also embrace Deming's mantra? In a word: arrogance. In the post–World War II global economy, America was king. The American manufacturing sector had a condescending approach to customers that is best epitomized by the sign that still hangs on a bridge crossing the Delaware River in Trenton, New Jersey: TRENTON MAKES, THE WORLD TAKES (see Figure 3.1).

Deming's total quality management mantra of planning, doing, checking, and acting fell on deaf ears in the United States in the 1950s. The American manufacturing approach at the time was: If a product breaks, send it back, and we will ship you another one.

Figure 3.1 "Trenton Makes, the World Takes"
Source: Ken Ficara

Datsuns Arrive in Los Angeles

The American approach worked throughout most of the 1950s. But in 1959, just nine years after Deming's Mount Hakone Conference Center speech, the Japanese manufacturing machine began making inroads when Datsun, who later changed its name to Nissan, began exhibiting cars at the 1959 Los Angeles Auto Show. Although Datsun did not have huge sales with the American car consumer, who still preferred big gas-guzzling automobiles, that all changed in the 1970s when the Arab oil embargo inflicted pain on the American economy. Detroit responded with poorly designed and manufactured cars like the Ford Pinto, which had a critical problem in the safety design of the gas tank. The smaller, fuel-efficient Japanese cars quickly won market share in the American automobile market.

Japan then continued this trend in the office equipment, photography, electronic entertainment, and steel industries.

For the first time, Japan was getting the attention of America's political leaders. On June 24, 1980, NBC News broadcast a 90-minute program that directly challenged the conventional wisdom of American politicians and business leaders, who seemed content to explain away Japan's success as a unique cultural process that could not be replicated in America.[6] But the NBC program, which featured Deming, proved them wrong. Japan's remarkable manufacturing resurgence in the 1960s and 1970s was largely a result of embracing the total quality management advice of one American mathematician, W. Edwards Deming.

This advice was ignored by American business and political leaders until the late 1980s.

American Business Leaders Finally Listen

Deming would remain disappointed that America never embraced his total quality management philosophy. He worked as a consultant, but after the NBC program brought his ideas to the front stage of American economic and political thought, he decided to put his theory in a book. In 1982 he published *Out of the Crisis*, in which he presented his total quality management advice to American business leaders on how to compete with the Japanese.

The preface to the book offered this blunt advice to American business leaders:

> The aim of this book is transformation of the style of American man-agement. Transformation of American style of management is not a job of reconstruction, nor is it revision. It requires a whole new

structure. Transformation must take place with directed effort. The aim of this book is to supply the direction."[7]

Four years later, in 1986, one American business, the Motorola Corporation, embraced Deming's total quality management principles when it introduced a manufacturing process called Six Sigma, which sought to improve the quality of process outputs by identifying and removing the causes of defects and minimizing variability in manufacturing and business processes.

This was exactly the advice that Deming had shared with Japan 36 years earlier.

A year later, in 1987, the U.S. Department of Commerce followed Motorola's lead, launching the Malcolm Baldrige National Quality Award to showcase performance excellence in private and public firms in the United States—a mere 37 years after the first Deming Prize was awarded in Japan.

Lessons from Deming

What is the key lesson about how an American skilled in science, technology, engineering, and math turned the global economy on its head? Actually, there are a couple of key lessons.

First, the American public school system helped a student skilled in science and math to reach his potential and change the global economy. That's good—although it wasn't good for the American firms that had to compete with Japanese companies in the 1960s, 1970s, and 1980s.

Second, America is a proud country, at times even arrogant. In 1950, Deming offered America's business leaders the same theories of total quality management that he shared with Japan's leaders, the same theories that the U.S. Army had used during the war. Japan embraced Deming's way; American business did not, even though the U.S. military had seen firsthand the power of Deming's total quality management methods during and after the war.

It took the United States nearly four decades to realize this critical mistake.

Deming's total quality management theory, which altered the global economy, is based on 14 principles that could dynamically change your business today:

1. Create constancy of purpose toward improvement.
2. Adapt to the new philosophy of the day; industries and economics are always changing.
3. Cease dependence on inspection to improve quality by building quality into the product in the first place.

4. End the practice of awarding business on basis of price tag.
5. Improve constantly and forever the system of production and service to improve quality/productivity.
6. Institute on-the-job training.
7. The aim of management should be to help people.
8. Drive out fear.
9. Break down barriers between departments.
10. Eliminate slogans, exhortations and targets for the workforce asking for zero defects and new levels of productivity. Such slogans only create adversarial relationships.
11. Eliminate quotas on the factory floor. Substitute leadership.
12. Remove barriers that rob the hourly worker of the right to pride of workmanship.
13. Institute a vigorous program of education and self-improvement.
14. Put everyone in the company to work to accomplish the transformation.[8]

Can Total Quality Management Fix the American Education System?

Here's an intriguing assignment. Reread the opening paragraph from this chapter. Wherever you see the word *management*, think government leaders, school administrators, teachers, and parents. When you come to the phrase *waste of manpower*, think of the 49.6 million students in America's public schools. For *consumer*, think of the American taxpayer and businesses that complain that workers do not have the skills needed to compete in the twenty-first century. For *materials*, think of the outdated school buildings and the boring 300-page textbooks that kids lug to school each day. And for *loss of market*, think of how the emerging regions of the world are rapidly overtaking the United States in many markets.

America's education problems are not "ordained," nor are they "inevitable." They are "man-made," and men and women in America can fix them.

I believe the ideas of W. Edwards Deming can help the American education system get back to the top of the global market. It will, however, take a long time, just as it took several decades to embrace Deming's concepts on the importance of total quality management.

Two years after Deming returned to the United States from his 1950 lecture tour, a new generation was about to enter the American education system, and this generation was about to be subjected to horrendous educational conditions. More than 60 years later, this generation remains in the position of influencing the quality of public education in the United States.

Notes

1. W. Edwards Deming, *Out of the Crisis* (Cambridge, MA: Massachusetts Institute of Technology, 1982), 9.
2. Robert B. Austenfeld Jr., *W. Edwards Deming: The Story of a Truly Remarkable Person* (Sacramento, CA: International Quality Foundation, 2001), 59.
3. Mark Magnier, "The 50 People Who Most Influenced This Century," *Los Angeles Times*, October 25, 1999.
4. John Hunter, "Speech by Edwards Deming to Japanese Leaders, 1950," Curious Cat, www .curiouscat.com/deming/deming-1950-Japan-speech-mt-hakone.cfm.
5. Ibid.
6. Lloyd Dobyns, "If Japan Can, Why Can't We?" NBC News, June 24, 1980.
7. Deming, *Out of the Crisis*, 9.
8. Ibid., 23–26.

CHAPTER 4

1952: Boomerang

Teachers hold in their hands the malleable minds of the nation's children. But despite the immense importance of what they do—or should do—they are wretchedly overworked, underpaid, and disregarded. And a discouraging number of them are incompetents.
—SLOAN WILSON, "CRISIS IN EDUCATION"[1]

As W. Edwards Deming applied his methods of statistical sampling to the Japanese postwar census, 7,500 miles to the east, another population event, the baby boomer explosion, was about to fundamentally reshape the American landscape in the 1950s—and affect the quality of math and science education in the United States into the twenty-first century.

The baby boomer population surge began in 1946 and continued until 1964. By the end of that 18-year period, the population of the United States had grown by an astounding 50,500,000. In the fall of 1952, when the first of the boomers turned six, 2 million of them arrived on the doorsteps of America's public schools. In 1953, another 2 million arrived. This population tsunami continued uninterrupted until 1970. America's public school system wasn't prepared for this onslaught.

What It Means to Teach

There was an "education war" raging in America in the early 1950s. An education war that was described by David Klein as "best understood as a protracted struggle between content and pedagogy."[2] Simply put, there was a contentious debate about how to teach and what to teach. And America's baby boomer schoolkids were about to enter its crosshairs.

A group known as progressives were led by John Dewey, a psychologist and education reformer. For this group, education was "active and schooling unnecessarily long and restrictive. Children came to school to do things and live in a community which gave them real, guided experiences which fostered their capacity to contribute to society."[3] For progressives, the *how* of teaching was more important than the *what*. They believed that students thrive best in an environment where they are allowed to experience and interact with the curriculum. Moreover, all students should have the opportunity to take part in their own learning. Progressives supported and created departments of education in American colleges and universities designed to produce what they believed were competent teachers who knew how to teach but had not mastered the subject matter to be taught.

On the other side of this debate were academic professors skilled in the content of a subject, like math. They believed that teachers earned their legitimacy to teach through years of subject mastery. The *how* of teaching would be learned through trial and error in the classroom. This group's leader was Arthur Bestor, a professor emeritus of history at the University of Washington in Seattle, who wrote the manifesto for this point of view in his 1953 best seller:

> A division into two educational worlds is threatened because the first twelve years of formal schooling in the United States are falling more and more completely under the policy-making control of a new breed of educator who has no real place in—who does not respect and who is not respected by—the world of scientists, scholars, and professional men.[4]

A Teacher Shortage Exacerbates the Educational Challenges

Although the debate over what to teach and who should teach continued, having a significant effect on public education for decades to come (remember the "open classroom" and "new math"), the debate was made irrelevant by other more powerful factors that were about to challenge millions of unsuspecting baby boomer students.

First, there was a shortage of 132,000 K–12 teachers in the United States.[5] Before this shortage, the certification process to become a teacher in America, though varying widely from state to state, had been rigorous. Not just anyone could show up and qualify to be a teacher. The shortage crisis presented those responsible for public education in America with a difficult choice: Should the teacher certification process remain committed to selecting only highly qualified teachers, or should it be modified to allow more applicants, many of whom would not have qualified to be a teacher in the past, to get teacher certification?

Pragmatism won out. America was in a bind. It needed teachers. Although the decision was never formally legislated, across the country, teacher certification requirements began to be lowered to the point where little professional training was considered necessary. Some states completely eliminated the certification process. Tens of thousands of Americans, mostly women, applied for easier-to-obtain teaching jobs. The teacher shortage crisis in the 1950s may have been averted short-term, but America never really thought through the long-term consequences of putting the responsibility to teach in the hands of hundreds of thousands of poorly trained teachers.

America continues to pay for these consequences in 2013.

Another Problem: Crumbling Infrastructure

What should be taught? Who should teach? Will *someone* please teach? Incredibly, these weren't the only challenges affecting baby boomer students.

Another challenge was horrible school infrastructure conditions. In 1952, the U.S. Office of Education said that America's schools were in a state of "shocking disorder and ineffectiveness."[6] In 1950 there were 166,473 elementary and secondary schools across America. These schools accommodated 29 million students, and the average school was rather small, with about 175 students. But the key problem was that tens of thousands of those schools were one-room school buildings built for education in the nineteenth century. From 1950 through 1958, just as the baby boomers began to arrive, many one-room schools were closed and absorbed into larger buildings.

So here's the picture: 16 million more kids were crammed into 135,372 schools, with the average school now serving 302 students. By 1953, the Office of Education reported a shortage of 345,000 classrooms.[7] This meant that 60 percent of American classrooms were overcrowded (see Figure 4.1). I can attest firsthand to this reality. In my first- through eighth-grade classes, I had an average of 60 kids in each class.

Figure 4.1   Close Quarters in an American Classroom
Source: Minnesota Historical Society

Media Critiques Begin

By the mid-1950s, the media had begun to focus on the plight of America's public schools. In 1955, Rudolf Flesch wrote a best seller entitled *Why Johnny Can't Read: And What You Can Do About It*, a critique of the then-popular method of teaching children to read by memorization.

However, the most significant critique of what was happening in American public schools came in March 1958, when *Life* magazine published a five-part series and dedicated two cover stories to address the topic—a highly unusual amount of coverage for any subject.

Life's editors said the following about the American education crisis:

> Teachers are too few and too hard-pressed in a nation-size job. Among the many problems of the public schools, the weakness in teaching is one of the most crucial. . . . And a discouraging number of them [the teachers] are incompetents.[8]

The March 24 cover story, which opened the series, featured two teenage boys, one from Chicago and the other one from the Soviet Union. The article focused on how they spent their time inside and outside the classroom. It was a chilling, damning comparison that showed American kids hanging out after school at dance parlors while their Soviet counterparts were studying late into the night.

Back in the USSR

Coming nearly six months after the Soviets had set off a firestorm of science and math education debates in America with the launch of *Sputnik* on October 4, 1957, the Soviet focus in the series' initial report played directly into America's fear that the Soviets were ahead of our country also in how we teach math and science. The White House thought the Soviets were doing a better job than Americans in teaching math and science, and in a highly unusual occurrence, particularly when you consider the escalating rhetoric of the Cold War at the time, President Eisenhower sent two delegations of education and business leaders to Moscow to learn how the Soviets were teaching math and science in their classrooms. Each delegation issued reports (see Figure 4.2, which shows the title page of one of these reports).

The first delegation chairman, L. G. Derthick, the U.S. commissioner of education, wrote the following about his group's observations of Soviet schools:

> We make no effort to compare the schools of the United States with those of the U.S.S.R., for we must measure the progress of each by its own separate goals. But we do emphasize that, whether we like it or not, competition has been imposed upon us by a nation of vast resources, a people of seemingly unbounded enthusiasm for self-development, governed by a ruling hierarchy which is determined to use that self-development to cast about the world the shadow of Communist domination.
>
> To sense this issue at first hand is indeed a sobering experience. We came back deeply concerned about our poorer schools now suffering from neglect. But we returned with a new appreciation and renewed faith in the American system, but the question is this: Will we Americans work and sacrifice to extend to all of our youth the best in American schools?[9]

Let's recap. As the largest population explosion in our country's history was descending on America's public schools, our nation was having a debate

Figure 4.2   Sample Report on Soviet Education

on what it meant to be a teacher, facing a critical shortage of people to teach, and lacking the proper infrastructure to house teachers and students across the country. The media started to write about our education inefficiencies, and our government was so lost on how to teach science and math that it reached out to our nation's most dreaded adversary for guidance. It was a mess.

That mess stretches into twenty-first-century America.

Boomers Perform Poorly on SATs

Most readers of this book took the SAT in high school. Owned, published, and directed by the College Entrance Examination Board, the SAT has been taken by millions of American students since 1926. In researching this book, I discovered a fascinating fact: from 1963 through 1976, the aggregate yearly SAT score for math dropped every year.[10]

Pardon the pun, but you do the math!

Since the baby boom generation began in 1946, and high school students generally take the SAT when they are 17, the students who took the 1963 SAT were the first class of baby boomer high school graduates.

There were 13 more classes of baby boomers, educated by poorly trained teachers, through 1976. Many of these teachers were not certified to teach; did not have degrees in science, technology, engineering, or math; and were forced to work in horrid working conditions.

Now let's shift our focus to the end of the twentieth century.

Trends in International Mathematics and Science Study (TIMSS) is a well-respected global assessment of the science, math, and reading skills of fourth-, eighth-, and twelfth-grade students. It was launched in 1995 and is conducted every four years. In every study—1995, 1999, 2003, and 2007—American students have performed poorly in math and science compared to students from other countries, a topic that will be addressed in depth later in the book. Moreover, in domestic math and science studies conducted by the Department of Education, more than 60 percent of American students are not grade-proficient in math or science.[11]

I have a hypothesis about this. Some people to whom I have spoken about my hypothesis think it may be a stretch, but others believe it has validity.

Connecting the Dots

Here is my attempt to connect the dots of the past 70 years.

The baby boomer explosion put incredible stress on America's education system in the 1950s through the mid-1960s. Many teachers in that time frame, though well-intentioned, were not prepared to teach, nor were they well trained in math and science. From 1963, the year the first baby boomer group took the SAT, through 1976, the aggregate SAT math score dropped consecutively each year. Obviously, the baby boomers did not learn math well from their second-rate teachers. American students' scores dropped precipitously in math and science from grades 4 through 10 in every international education assessment test since 1995. Most schools in America, according to the stringent standards of the Department of Education, are failing to teach math and science in a grade-proficient manner. Today, 51 percent of America's teachers are still baby boomers, on the verge of retirement.[12] Thus, all of our teachers—including those who were students of the baby boomers—were taught in less than ideal conditions by less than competent teachers. In the

2011–2012 school year, the United States ranked fifty-first in the World Economic Forum's list of countries in "quality of math and science instruction."[13]

The Boomerang Theory

I call my hypothesis the boomerang theory.

The turbulent education environment of the 1950s resulted in hundreds of thousands of women and men becoming teachers who really should not have been accepted into the profession. The baby boomer students were never given the opportunity for a quality education provided by quality teachers in adequately equipped schools and classrooms. The 14-year consecutive drop in SAT math scores that began in 1963 was not coincidental; it began just as the first wave of baby boomer students took the exam. Today, 51 percent of America's teachers are baby boomers, who lost their opportunity for a quality public education in the 1950s and 1960s and can never get it back. According to my theory, the ebbing performance of American students in international math and science tests since 1995, and the proficiency results of the National Assessment of Educational Progress (to be discussed later), is due, in part, to the teaching inadequacies of baby boomer teachers.

In May 1951, Sylvia F. Porter, a columnist for the *New York Post*, wrote a column that coined the term *baby boomer*. She said, "Take the 3,548,000 babies born in 1950. Bundle them into a batch, bounce them all over the bountiful land that is America. What do you get? Boom. The biggest, boomiest boom ever known in history."[14]

Baby boomers were the unfortunate victims of a public education system that was unprepared for them. In the past 67 years they have bounced "all over the bountiful land that is America" and further affected the quality of public education in the United States.

Their own education made it harder for them to inspire subsequent generations of Americans to understand and excel in math and science.

Notes

1. Sloan Wilson, "Crisis in Education," *Life*, March 24, 1958.
2. David Klein, *A Brief History of American K–12 Mathematics Education in the 20th Century* (Charlotte, NC: Information Age, 2003), 3.
3. James Neill, *John Dewey: Philosophy of Education*, Wilderdom, www.wilderdom.com.
4. Arthur Bestor, *Educational Wastelands: The Retreat from Learning in Our Public Schools* (Chicago: University of Illinois Press, 1985), 292.

5. "The 1950s: Education Overview," Encyclopedia.com, http://www.encyclopedia.com/doc/1G2-3468301830.html.
6. Ibid.
7. Ibid.
8. Wilson, "Crisis in Education."
9. L. G. Derthick, *Soviet Commitment to Education: Report of the First Official U.S. Education Mission to the U.S.S.R.* (Washington, DC: U.S. Office of Education, 1958).
10. College Entrance Examination Board, *On Further Examination: The Report of the Advisory Panel on the Scholastic Aptitude Test Score Decline* (New York: College Entrance Examination Board 1977), 13.
11. National Center for Educational Statistics, *Trends in International Mathematics and Science Study* (Washington, DC: U.S. Department of Education, 2011).
12. Jeanette Der Bedrosian, "A Tsunami of Boomer Teacher Retirements Is on the Horizon," *USA Today*, April 6, 2009.
13. World Economic Forum, *Global Competitiveness Report, 2011–2012* (Geneva, Switzerland: World Economic Forum, 2012), 363.
14. Sylvia Porter, "Sylvia Porter" (column), *New York Post*, May 1951.

CHAPTER 5
1962: Too Hard to Follow

61% of U.S. middle school students would rather take out the trash than do math homework.

—RAYTHEON CORPORATION[1]

On September 12, 1962, President Kennedy traveled to Rice University in Houston, Texas, to note the one-year anniversary of his administration's decision to build the National Aeronautics and Space Administration's "Manned Spacecraft Center" in that city. It was also Kennedy's second speech on his vision of a manned flight to the moon by the end of the decade.

Kennedy's first call to action on this goal didn't come in a speech on this bold initiative. Rather, it was a 173-word paragraph buried in a 5,856-word speech to Congress entitled "Special Message to Congress on Urgent National Needs," delivered on May 25, 1961. Here the President had to explain, among other topics, the embarrassing failure of the Bay of Pigs invasion a month earlier.

Houston was different. The goal there was nothing other than to promote the importance of the moon mission, and the growing importance of NASA, in Vice President Lyndon Johnson's home state two months before the midterm election.

With a battery of television cameras placed near the podium, President Kennedy delivered a masterful speech (you can watch it on YouTube) on a hot and sultry day challenging the country to land a man on the moon, and return him safely, by the end of the decade.

The Rationale for the Lunar Landing

Why President Kennedy chose the end of the decade as the goal for the moon mission has always puzzled me. My assumption was that his advisors instructed him to place the completion of the mission outside the parameters of his potential eight-year presidency, which would have ended on January 20, 1969. If the mission were successful, he still would get credit. If it hadn't happened by December 1969, it would have been the next president's fault.

That wasn't the reason, however; pragmatism was. America was woefully behind the Soviet science, technology, engineering, and math (STEM) initiatives. Although Wernher von Braun and his team of German scientists at NASA continued to make progress on rocket experimentation through the early 1960s, the Soviets were far ahead of America for two key reasons.

First, their structured approach to teaching STEM was producing more scientists and engineers faster than the American education system, which was still coping with the influx of baby boomers. The second reason was organizational. Throughout the 1950s and early 1960s, the Soviets had centralized their rocket development efforts. The United States, on the other hand, was only beginning to learn the advantages of centralized research and development under NASA after a decade in which rocket development had been a free-for-all competition among the army, the navy, the air force, and civilians.

Therefore, the only realistic goal that America *might* be able to beat the Soviets at was the ambition, seven years down the road, to send a man to the moon. Kennedy knew this reality and acknowledged America's second-place position in the space race at the beginning of his speech.

Kennedy in His Own Words

We have had our failures in manned flight, but so have others. To be sure, we are behind, and will be behind for some time in manned flight. But we do not intend to stay behind, and in this decade, we shall make up and move ahead. But why, say some, the moon? Why choose this as our goal? And they may well ask why climb the highest mountain? Why, 35 years ago, fly the Atlantic? Why does Rice play [the University of] Texas? We choose to go to the moon in this decade and do the other things, not because they are easy, but because they are hard, because that goal will serve to organize and measure the best of our energies and skills.

—JOHN F. KENNEDY, RICE UNIVERSITY ADDRESS[2]

I almost called this chapter "And Do the Other Things" because I was going talk about all the IT leaders who were influenced by Kennedy's 1962 speech at Rice University about America's mission to the moon: Vint Cerf, who invented key components of the Internet; Bill Gates and Steve Jobs, who were both seven years old at the time; Gordon Moore, who was a cofounder of Intel; and Robert Metcalfe, who invented the local area network and was a teenager at the time. These are people who did the "other things." But I changed my mind, because the technology hit parade, while interesting, was mundane.

Kennedy's comment "because they are hard" is what hit the mark.

"It's Just So Darn Hard"

Kennedy's comment that we would go to the moon because it was hard, not because it was easy, meshes well with a 2011 article in the *New York Times* that focused on college students opting out of studying math and science because those subjects were perceived as "just so darn hard." Studies quoted American students saying, "I hate math just because it's hard for me to understand" and "Science doesn't matter unless you want to become a doctor or something like that."[3]

Why do Americans continue to harbor the perception that math and science are hard subjects to learn? According to the Public Agenda Foundation, this perception begins to form in the home. In 2007, with support from the Ewing and Marion Kauffman Foundation, the Public Agenda Foundation studied 2,767 parents and students in Kansas and Missouri. Among the most prominent findings are these:

- 57 percent of parents say that the United States is far behind other countries in math and science achievement.
- 10 percent say that America is well ahead of the rest of the world.
- 86 percent agree that students with advanced math and science skills will have a big advantage in work and college opportunities.[4]

These findings—particularly the one that nearly 6 in 10 parents think America is "far behind other countries in math and science achievement"—would suggest that parents are looking for change in math and science classrooms.

Yet when the parents were asked if their child's school should be teaching "a lot more math and science, less math and science, or are things just fine as they are," 70 percent opted for the things-are-just-fine response. That seems disconnected from the earlier answer about America's competitiveness position. The answer to another question in the survey tells why: 64 percent said that "schools have more basic problems to solve before they can start worrying about improving math, science, and technology education."[5]

So while parents in the study recognize America's eroding position relative to the rest of the world in science and math education *and* appreciate the importance of those subjects to work and career pursuits, they nevertheless seem content to kick the let's-improve-math-and-science-education can down the road until "more basic problems" (not alluded to in the study) are solved.

Students: Math and Science Are Irrelevant

The Public Agenda Foundation study also discovered that math and science were perceived by the students themselves as irrelevant to their lives: 76 percent claimed that "poor achievement in math and science can be chalked up to the fact that [the students] find these subjects irrelevant," and 50 percent of the parents agreed with their children. Moreover, when asked about enrolling in high-level math and science courses, only 27 percent of students said that all students should be required to take them, whereas 72 percent said that only students who are "interested" should take them.[6]

The Public Agenda Foundation study was not done in isolation. In December 2010, the Raytheon Corporation released the results of a global study that aimed to determine the connection between parental attitudes toward math education in America, England, and Singapore. The four prominent findings of the study were as follows:

1. Parents in all of the regions lacked a clear understanding of their children's math abilities, but the parents in Singapore are were twice as likely as the others to ensure that their children received extra math instruction beyond the classroom.
2. While the U.S. parents claimed to be more confident in their ability to help their children in all areas of math, once middle school math was introduced, that confidence waned.
3. Five times as many parents in Singapore than in the United States reported that they receive information on their children's math performance from the school.
4. Whereas U.S. students are more likely to participate in spelling competitions, students in Singapore are three times more likely to participate in math competitions.[7]

Culture Counts

Seymour Papert, the founder of the Massachusetts Institute of Technology (MIT)'s Media Lab, wrote about schools and what children learn in his oral history filed with the Computerworld and Smithsonian Institution archive in Washington, D.C. He said the following:

> Much of what children learn is not actually taught. When you look at school and you see how they are teaching different things, you ask why is it that algebra needs such enormous efforts, whereas things that children learn, like language and talking and logic and getting around a space and all the rest, seems to happen so naturally? The answer is that children will easily learn intellectual material, they will construct intellectual material, if they've got the materials for constructing it in their education. Schools just can't do that. They are trying to change schools by having more effective teaching. It just doesn't work. It's only when it's part of the culture and is in their lives that it can really become part of that learning, which happens in this natural way.[8]

Bill Gates, the cofounder of the Microsoft Corporation and the chief executive officer (CEO) of the Bill and Melinda Gates Foundation, amplified Papert's points in a panel discussion at the 2010 Techonomy Conference when he said that our education system must "let students assess their knowledge, let them know where they are, let them proceed at a pace where they know what they know before they move on."[9]

Some might think that Papert's ideas are a bit too idealistic. The reality, for better or worse, is that the perception of math and science as hard subjects only becomes stronger in middle school, particularly for girls. At this level, 69 percent of math classes and 57 percent of science classes are taught by teachers who are not mathematicians and scientists—that is, educators who have no undergraduate, graduate, or doctoral degree in science or math.[10]

Industry Leaders Offer Advice

Why is it important that math and science are being taught by teachers who have no degrees in these subjects? Paul Otellini, the former chairman and CEO of the Intel Corporation, explains why:

> Getting a child interested in STEM requires inspiration. In the unlikely case that the child's father or mother is an engineer, there is a natural role model. Barring that, without an inspiring teacher, there is no one to role model. Add that to the fact that math and science are harder than other subjects and you get "irrelevance."[11]

Bill Gates, at the Techonomy Conference in 2010, took direct aim at math and science textbooks of 300 pages or more in middle school and high school. He called them "giant, intimidating books. I look at them and think, what on earth is in there? In fact, our textbooks are three times longer than their equivalents in Asia. The problem with American textbooks is that they are built by committee, and more things are simply added on top of what's already in there."[12]

Do Something about It

Richard Miller, president of the Franklin W. Olin College of Engineering in Needham, Massachusetts, has claimed that math and science are considered hard subjects to learn because "there is too much emphasis in the classroom placed on processing a body of knowledge than [on] application of that knowledge by doing. We need to teach our students from middle school through higher education to 'do' rather than 'know.'"[13]

Michael Gabriel, chief information officer (CIO) at HBO, shares Miller's view. He said that math and science are perceived as "irrelevant because most teaching relies too much on the textbook and rarely shows a correlation to how the subject can actually either influence their career or their life; without any context behind it, the teaching is a waste of time."[14]

Robert Herbold, the managing director of the Herbold Group and a former CIO and chief operating officer (COO) at the Microsoft Corporation, pointed his finger at the teacher unions and "their stonewalling of the installation of performance appraisals and termination for poor performance, but given their $300,000,000 per year of political clout at the local and national level, don't hold your breath" for reforms.[15]

Miller shared with me a personal story of doing rather than knowing in higher education:

> In the 1950s, 90 percent of engineering professors at the undergraduate level had no PhD degree in the engineering field they were teaching. They earned their way into the college classroom by *doing* things. In 2012, it is entirely reversed. Ninety percent of college-level engineering professors have PhDs and are more focused on basic research that will lead, hopefully, to a tenured position within ten years rather than engaging students with an innovative, imaginative curriculum focused on experiential learning. How engineering is taught across America is like going to a music college where the curriculum focuses on the instrument rather than encouraging the student to write and play music.

He added, "We must teach science and math like it were a Nancy Drew mystery story."[16]

Steven Hoover, the president of Xerox Palo Alto Research Center, believes the American education system needs to make science and math education "fun."[17]

American Students Not Measuring Up

Whether math and science are perceived as hard, irrelevant, mysterious, or fun, students have no choice. They must take basic math and science courses. State and national laws like the current version of the Elementary and Secondary Education Act of 1965, commonly known as the No Child Left Behind Act, mandate that students take math and science from grade three through high school. The act also states that the students be tested regularly on how well they have learned math and science (and other subjects) with an examination known as the National Assessment of Educational Progress (NAEP), created by the U.S. Department of Education, pilloried throughout America as "the test" and giving rise to the phrase "teaching to the test."

The NAEP reports student test results in four groups:

1. **Below basic.** Has not mastered the subject matter.
2. **Basic.** Has demonstrated a partial mastery of the subject matter.
3. **Proficient.** Has demonstrated a mastery of the subject matter.
4. **Advanced.** Has demonstrated a superior mastery of difficult subject matter.[18]

When the No Child Left Behind Act was signed into law in 2002, it mandated to each state that in order to continue to receive federal funding for education, 90 percent of the students in the state must be rated as proficient or advanced in math and science by 2014.

The Results, Please

So how are America's kids doing in math and science? Not well. Table 5.1 shows the 2011 NAEP scores for math and science proficiency by grade level.[19]

For a nation that invests $10,441 annually per student, those are hard numbers to accept. And perhaps most troubling is that the lack of proficiency only increases as the students move on in school.

Table 5.1 Not Very Proficient

Grade Level	Not Proficient In Math	Not Proficient In Science
Grade 4	60%	66%
Grade 8	65%	70%
Grade 12	74%	79%

How to Do Something

What can we do? Paula Golden, the executive director of the Broadcom Foundation, a private nonprofit organization focused on STEM issues, has a suggestion that is in sync with Richard Miller's focus on doing. Golden wrote an article in 2011 in which she said that "the American education system, and parents, should expose kids to the real-world applications of science and math; if a middle school student is interested in texting, let them [the kids] open up the phone and explore its parts."[20]

The Broadcom Foundation partners with the Society for Science and the Public to produce a program called the Broadcom MASTERS (which stands for Math, Applied Science, Technology, and Engineering as Rising Stars) which honors sixth-, seventh-, and eighth-grade students who display exceptional skill in the mastery of STEM.

Irving Wladawsky-Berger, the former chairman of IBM's Academy of Technology and now a professor of IT at MIT, seems to agree with Golden's real-world application approach to teaching science and math. He said, "The vast majority of people learn best through concrete cases and experience, through lots and lots of examples and hands-on experiences."[21]

When Richard Miller was an engineering professor at the University of Southern California (USC), he was frustrated trying to bring real-world applications of math and science to the Pasadena School district, which is near the USC campus. He reached out to Pasadena school administrators with the idea that he and other engineering professors would volunteer time to visit schools in the district and teach elective courses like computer science.

Their overture was promptly declined for two reasons: (1) even though Miller and his college-level faculty were qualified to teach all aspects of engineering to students at USC, they were unqualified to teach in the Pasadena school district's middle schools or high schools because they did not have a state-approved certification license to teach at those levels, and (2) if the Pasadena school district were to offer elective courses in computer science, the current faculty would have first dibs at teaching those courses. This was the case because elective courses were considered easier to teach because only motivated students would sign up for them, and the certified, unionized faculty wanted those plum courses for themselves.

What Miller and his colleagues were offering in 1980 were their services as adjunct teachers, aspiring to complement the efforts of the classroom teachers with supplemental knowledge. Several weeks after visiting with Miller in April 2012, I read an article in a Worcester, Massachusetts, newspaper about how the city of Worcester turned down an offer by Teach for America, a national program that sends high-achieving college graduates as volunteers to high-poverty classrooms, because "the participants did not have enough preparation for the first time they lead a class, and having these volunteers in Worcester's classrooms would do the 24,000 students in Worcester a disservice."[22]

High School Seniors: No, Thank You

After seven years of compulsory math and science instruction in middle school and high school, after seven years of sitting through repetitive and unimaginative "teaching to the test" lesson plans, it should come as no surprise that when American high school seniors are deciding what degrees they wish to pursue in college, they run the other way from STEM.

So says a January 2012 report from Business Higher Education Forum, the nation's oldest organization of senior business and higher education executives. The report analyzed a 2008 data set of Missouri twelfth-grade students for their level of interest in pursuing STEM majors in college and then compared those interest levels to their math proficiency scores in the American College Testing (ACT) exam. Here's what the study found:

- 17 percent of the seniors were "math proficient and interested" in STEM degrees in college.
- 14 percent of the seniors were "not math proficient but interested" in STEM degrees in college.
- 27 percent of the seniors were "math proficient but not interested" in STEM degrees in college.
- 42 percent of the seniors were "not math proficient and not interested" in STEM degrees in college.

That is, 83 percent of the seniors were not planning to pursue a STEM degree because they were either "not proficient" or "not interested."[23]

A further analysis of these numbers indicates that after 12 years of instruction, 56 percent of the students were still not proficient in math skills, so it is understandable that they would look to degree programs in other subject areas. More than 60 percent of the seniors deemed "math proficient" said "no, thank you" to studying STEM in college. But the most damning fact of all is that 69 percent of these Missouri seniors said they were not even interested in STEM degrees.

Why is there such a lack of interest among high school seniors in undergraduate degree programs in STEM?

David E. Goldberg, a retired professor of engineering at the University of Illinois, Urbana, has said he knows why: science and math are just "dry and hard to get through."[24] Paula Golden doesn't buy this argument, however. She says that "no matter what direction kids choose to go in their careers—banking, building homes, designing microchips—they need math and science. The simple truth is math and science aren't any harder than social studies or history. They take practice and discipline. It is a matter of once again making the subjects relevant to children, especially in middle schools."[25]

Perception Is Reality: The Importance of the Guidance Counselor

Although national attention mostly focuses on uninspiring teaching methods, such as "teaching to the test," and unimaginative curriculum in the classroom as the primary reasons American students shun careers in STEM, I believe that another member of the education community contributes to cementing the impression that math and science are hard subjects to master: the middle school and high school guidance counselor.

I have harbored this impression for years. My two children attended middle school and high school at our local public schools during the dot-com years, when I was the publisher of *Computerworld*. In conversations with members of the schools' guidance departments during that time, I often came away bewildered at how unprepared the staff was to discuss careers in engineering, math, or science—and they had no idea what an IT worker did for a living.

The Stevens Institute of Technology, a respected technology college in Hoboken, New Jersey, decided to find out just how STEM-savvy guidance counselors are. The premise of the study was that if the counselors are naive about STEM careers, they are probably encouraging, or at the very least not refuting, the perception that math and science are hard subjects to master.

In 2008 the institute studied a sample of New Jersey middle school and high school guidance counselors to determine how they spend their time and what their depth of knowledge is about STEM subjects and careers in the field.

How you spend your time defines your job. Here's how the guidance counselors in the Stevens study spent theirs:

- College advising, 23 percent
- Scheduling, 22 percent

- High school advising, 14 percent
- Counseling, 11 percent
- Career advising, 9 percent
- Disciplining, 2 percent

This sample of guidance counselors responded to the Stevens survey after attending a one-day seminar on STEM careers. This is what they said *before* they attended the seminar:

- 85 percent did not know about job opportunities in STEM.
- 15 percent did not know what skills were needed for STEM careers.[26]

The respondents to the Stevens survey were also asked where they turned when they needed to find out more about careers in science, math, and engineering. Three primary sources were listed: (1) teachers, (2) professional development workshops, and (3) the media (Internet, TV, newspapers, and magazines).

The report offered this advice about the role of guidance counselors and STEM fields:

> Guidance counselors are a primary source of career information for students and have the unique opportunity to increase awareness about STEM careers with teachers, administrators, students, and parents. To improve guidance counselors' ability to effectively introduce and encourage students towards STEM careers, guidance counselors report that they require access to well-designed professional development workshops and access to role models, specifically undergraduate and graduate engineering students and professionals.

The conclusion of the Stevens study was that "the majority of K–12 guidance counselors hold misconceptions, or suffer from a lack of information about, STEM and STEM careers."[27]

Can you imagine the positive effect it would have on the STEM pipeline if all the readers of this book visited their local schools' guidance counselors to offer advice on what it means to be an IT executive in the twenty-first century and to share information on the positive employment trends in the IT field?

The STEM Pipeline Shrinks More in Higher Education

It is unfortunate that the challenge of filling America's future STEM employment needs begins only in middle school or high school. The drop-off in STEM education accelerates in higher education among students who are proficient in science and math and initially interested in getting degrees in those areas. The road to earning an undergraduate STEM degree has been called a "math and science death march," particularly in freshman and sophomore years, when the curriculum is dull, boring, and often more intent on weeding out weaker students than fostering a collaborative, participatory learning environment.

Nancy M. Hewitt and Elaine Seymour, then sociologists at the University of Colorado in Boulder, studied 460 college students enrolled in STEM degree programs and published the findings of their study in a book in 1997. They list the top 10 reasons that undergraduates enrolled in STEM programs switch to other degree disciplines (see Table 5.2).[28]

Other reasons cited by Hewitt and Seymour include STEM faculty, who are "often represented as 'unapproachable' or 'unavailable,'" "harsh grading systems which discourage collaborative learning strategies," and "teaching assistants bearing a disproportionate responsibility for the teaching of fundamental classes that are often over-enrolled." Other forces that accelerate the STEM dropouts in college are the specter of grade inflation in degree programs like liberal arts.

A *New York Times* article in 2011 published 1,110 comments from readers and underscored the battle going on in universities with STEM education.

One submission, seemingly from a professor, said, "I've concluded that unless a student is truly interested in science and has good self-discipline and is willing to work very hard, [he or she] will not succeed in one of these disciplines. It is very rewarding if you stick with it and master the basics and learn to apply the principles, but getting to an intuitive understanding of nature and physical systems takes years of hard work, too. It seems many just won't make the effort."[29]

Table 5.2   I Am Out of Here!

Factor	Percent Citing
Lack of interest in STEM	43%
Other degrees more interesting	40%
Poor teaching methods	36%
Curriculum overload overwhelming	35%
STEM career prospects not worth the effort	31%
Rejection of STEM careers and lifestyles	29%
Shift to more appealing non-STEM career options	27%
Inadequate advisors	24%
Loss of confidence because of low grades in early years	23%
Inadequate finances to complete degree	17%

A student who had hoped to become an astrophysicist said, "When I enrolled in college I took an introductory astronomy class—and it was the most boring, tedious thing I ever endured. I thought we would be observing the heavens through telescopes, not slogging through obscure information about celestial spheres. I decided that while science is a fun hobby, I didn't want to do something this boring for a living and so I switched to studio art."[30]

Putting Words in the President's Mouth

Let's end where we began this chapter: with President Kennedy's 1962 speech at Rice University. Only this time let's substitute *math and science* for *manned flight* and *the moon*:

> We have had our failures in math and science, but so have others. To be sure, we are behind, and will be for some time in math and science. But we do not intend to stay behind, and in this decade, we shall make up and move ahead. But why, say some, math and science? Why choose this as our goal? And they may well ask why climb the highest mountain? Why, 35 years ago, fly the Atlantic? Why does Rice play [the University of] Texas? We choose to [do] math and science in this decade and do the other things, not because they are easy, but because they are hard, because that goal will serve to organize and measure the best of our energies and skills.

But an essay written by an obscure Japanese physicist the same year President Kennedy delivered his speech argued that history would make it very difficult for America to win that fight.

Notes

1. Raytheon Corporation, "U.S. Middle School Students Math Study Habits," Math Moves U, http://www.raytheon.com/newsroom/rtnwcm/groups/corporate/documents/content/rtn12_studentsmth_results.pdf.
2. John F. Kennedy, "Address at Rice University on the National Space Effort," September 12, 1962, John F. Kennedy Presidential Library and Museum, www.jfklibrary.org.
3. Christopher Drew, "Why Science Majors Change Their Minds (It's Just So Darn Hard)," *New York Times*, November 4, 2011.
4. Will Friedman and Alison Kadlec, *Important but Not for Me: Parents and Students in Kansas and Missouri Talk about Math, Science, and Technology Education*, Public Agenda Foundation, 2007, www.publicagenda.org.
5. Ibid.
6. Ibid.

7. Raytheon Corporation, "Global Study Analyzes Parents' Math Attitudes, Capabilities and Students Active Engagement in Math-Related Learning," http://raytheon.mediaroom.com/index.php?s=43&item=1716, 2010.

8. Seymour Papert, Oral History Collection, Computerworld Honors Program, www.cwhonors.org/archives/histories.htm.

9. "In Five Years the Best Education Will Come from the Web: A Conversation with Bill Gates," Techonomy Conference, August 6, 2010, Tech Crunch, http://techcrunch.com.

10. National Center for Educational Statistics, *Out-of-Field Teaching and Educational Equality*, October 15, 1996, http://nces.ed.gov/pubsearch/pubsinfo.asp?pubid=96040.

11. Paul Otellini, e-mail to author, March 29, 2012.

12. "In Five Years the Best Education."

13. Richard Miller, interview with author, Needham, MA, March 28, 2012.

14. Michael Gabriel, e-mail to author, March 29, 2012.

15. Robert Herbold, e-mail to author, May 13, 2012.

16. Miller interview.

17. Steven Hoover, telephone interview with author, November 28, 2011.

18. National Assessment for Educational Progress, *Science 2011: National Assessment for Educational Progress at Grade 8* (Washington, DC: U.S. Department of Education, 2011), 13, http://nces.ed.gov/nationsreportcard/pdf/main2011/2012465.pdf.

19. Ibid.

20. Paula Golden, "Keep Science and Math Alive in Middle School," *Huffington Post*, October 4, 2011.

21. Irving Wladawsky-Berger, e-mail to author, May 2, 2012.

22. Jacqueline Reis, "Worcester Schools Look at Teacher Recruitment," *Telegram-Gazette*, April 4, 2012.

23. Business Higher Education Forum Research Brief, *Addressing the STEM Workforce Challenge: Missouri* (Washington, DC: Business Higher Education Forum, 2012), 1, www.bhef.com/publications/documents/BHEF_Research_Brief-Addressing_the_STEM_Workforce_Challenge-Missouri.pdf.

24. Drew, "Why Science Majors Change Their Minds."

25. Golden, "Keep Science and Math Alive."

26. Dawna Schultz, *Engineering Our Future New Jersey: Guidance Counselors Mission Critical* (Hoboken, NJ: Stevens Institute of Technology, 2008).

27. Ibid.

28. Nancy Hewitt and Elaine Seymour, *Talking about Leaving: Why Undergraduates Leave the Sciences* (New York: Alfred P. Sloan Foundation, 1997), 33.

29. Drew, "Why Science Majors Change Their Minds."

30. Ibid.

CHAPTER 6

1962: Empires of the Mind

The empires of the future are empires of the mind.
—WINSTON CHURCHILL, HARVARD UNIVERSITY, 1943[1]

Antoine van Agtmael, the author of *The Emerging Markets Century: How a New Breed of World-Class Companies Is Overtaking the World* (2007), disagrees with the basic premise of Thomas Friedman's 2005 best seller, *The World Is Flat*. In a March 2007 article he wrote, "The world is not flat; it is tilting, with the USA rapidly moving from unquestioned dominance to great dependence."[2]

Ten months later, the *Financial Times* published an article that asserted, "Although the U.S. remains a powerful business force, it can no longer take for granted its dominance over the rest of the world."[3]

Less than a month after that, Bob Suh, the former chief technology officer (CTO) of the global consulting firm Accenture and now the CEO of OnCorps, wrote in the same newspaper that American companies are falling behind in technology because they "dedicate the majority of their fresh capital to fortifying older systems while companies in Europe and Asia invest in more up-to-date systems."[4]

Did You Know?

It was in this environment that Karl Fisch, the director of technology at Arapahoe High School in Littleton, Colorado, was asked by school administrators to speak at the 2006 opening faculty meeting. The administrators wanted Fisch to share with the faculty the latest and greatest technology tools that they might use in their teaching efforts.

Fisch had a better idea. Rather than subject his colleagues to a mundane presentation on cool tech gadgets, he opted for a presentation that had relatively little to do with technology and everything to do with thought-provoking ideas on the fast-changing world that awaits Arapahoe High School students upon graduation.

His presentation, entitled "Did You Know: Shift Happens," was a huge hit at the meeting. It's on YouTube, where 5,540,000 people have viewed it. Here's an excerpt from the slide show:

Name this country . . .
Richest in the world.
Largest military.
Center of world business and finance.
Strongest education system.
World center of innovation and invention.
Currency the world standard of value.
Highest standard of living.
England.
1900.[5]

It grabs your attention, doesn't it? Try it on your friends, only stop after saying "highest standard of living" and ask them to guess which country the sequence is all about. Most people say the United States. All are shocked when you tell them it's England at the turn of the twentieth century.

The Shift Is On

Several days after viewing Fisch's video on YouTube, I received an e-mail from the Rand Corporation about a report on American competitiveness in science and technology. With Fisch's slide show fresh in my mind, a sentence in the opening paragraph of the Rand report intrigued me:

By one estimate, from the 16th century to the present, scientific centers in the West have shifted, with an average period of scientific prosperity of about 80 years (Yuasa, 1962).[6]

After I spent 30 minutes searching for Yuasa on the Internet, I learned that Mitsutomo Yuasa is a Japanese physicist who in 1962—when President Kennedy was encouraging Americans to look forward eight years to a moon landing—was looking in the opposite direction: back 473 years, to 1540.

What intrigues me most about Yuasa is how little is known about him. Aside from several articles in the *Boston Globe* and *CIO* magazine, the search returns are sparse.

But if you are persistent in your search efforts, you will find two scholarly articles written about Yuasa and his unique theory that the world's scientific center shifts from one country to another on a regular basis. In 1985, two Chinese scientists, Zhao Hongzhou and Jiang Guohua, reviewed Yuasa's work in the article "Shifting of World's Scientific Center and Scientists' Social Ages" in the journal *Scientometrics*. Fifteen years later, three other Chinese scientists, Liming Liang, Ye Feng, and Yishan Wu, also wrote about Yuasa's theory in the article "Shifts in the World Science Centre" in *Interdisciplinary Science Review*. (It should come as no surprise that these reviews of Yuasa's theory, the only reviews I could find, hail from the People's Republic of China, because many Chinese believe that it is to their country that the world's center of scientific activity will shift sometime during the twenty-first century.)

Yuasa's own 1962 article laid out this succinct theory: Since 1540, every 80 to 100 years the center of scientific activity in the Western world has shifted from one country to another.[7]

The Components of Yuasa's Phenomenon

Yuasa went back to the sixteenth-century Renaissance days of Leonardo da Vinci and focused his work on world scientific centers. In Yuasa's theory, which after years of peer-based scrutiny is now called Yuasa's Phenomenon, a country becomes a world scientific center when its scientific achievements exceed 25 percent or more of the world's most significant scientific achievements in a particular period.[8]

World scientific centers, according to Yuasa, always have a concentration of prominent scientists. A key exception to his theory is that even though country A is noted as the world's overall scientific leader, country B could be a world leader in a specific scientific discipline (but not necessarily the world's overall top scientific center).

Age was a key component of Yuasa's theory. He claimed that a country's scientific and technological leadership will decline if the average age of the country's outstanding scientists is more than 50 years old.[9] A quality education system that advances young students into scientific careers was another key component.

Fast-Forward

When I was the publisher of *Computerworld*, I had the honor of hosting for six years the annual Computerworld and Smithsonian Institution's Search for New Heroes awards ceremony, created in 1989 by International Data Group chairman Patrick J. McGovern and Museum of American History director Spencer Crew. Vint Cerf, the father of TCP/IP, the backbone of the Internet, was an award winner in 1996.

Several months after receiving his award, winners like Cerf were required to sit down with the curators of the Smithsonian Institution to file an oral history of their lives in technology. I recommend that every reader log on to the Computerworld Honors web site, www.cwhonors.org, and read the fascinating personal stories of Steve Jobs, Bill Gates, and others. It is the IT industry's most meaningful history trove.

In telling his oral history, Cerf shares with curator David Allison that in elementary school he was fascinated with algebra. Cerf had literally memorized every equation in his fifth-grade algebra book. Frustrated, he approached his fifth-grade teacher, Mr. Tomazewski, and asked him what he could do next. Tomazewski promptly handed the young Cerf a seventh-grade algebra textbook. Cerf was ecstatic.

The following year, 1997, Seymour Papert, the founder of the world-famous MIT Media Lab, was an award winner at the ceremony, and something he said in his acceptance speech illustrated why Tomazewski's decision to give Cerf a seventh-grade algebra textbook hit the mark. In the preface of this book I told how he pointed out the absurdity of promoting by age.

In his oral history filed with the Smithsonian Institute, Papert amplified his point:

> The most dramatic thing you would see if you could go in a time machine to a school of the future is there will be no grade [level]s. This idea of segregating children by age makes no sense except as an organizational means to hand out their curriculum information bit by bit in a systematic way. There is no justification for it. Nowhere else in life do we think it necessary to segregate people by age. We do this in school because we think that's the only way children could get access to knowledge. We are moving rapidly into a time when you can get knowledge when you need it, and can get it for use rather than for storing it away in some banking system in your mind. This is the way it is going to be. The idea of curriculum and the idea of grade level that goes with it will be thrown away.[10]

Yuasa's Phenomenon Arrives in America in 1920

Here's how Yuasa saw the shifting of the world's scientific centers from 1540 to 1920:

- Italy (Florence, Venice, Padua): 1540–1610
- England (London): 1660–1730

- France (Paris): 1770–1830
- Germany (Berlin): 1810–1920
- United States (New England and California): 1920–present[11]

If Yuasa's Phenomenon is in play again, there should be signs of a shift from the United States to another country by 2030.

I was curious, so I did my own review of Yuasa's Phenomenon. For significant scientific achievements, I decided to compile and analyze the nationalities of Nobel Prize award winners in four disciplines—economics, physics, medicine, and chemistry—from 1983 through 2011. (I purposefully chose 1983 because in that year the highly provocative report *A Nation at Risk* was released by the Department of Education. This report will be discussed later in the book.)

My premise was simple. I wanted an analysis by nation, from 1983 to 2011, to learn whether America's percentage of winners was increasing, holding strong, or giving way to another country. Was the United States maintaining a base of 25 percent of the world's most prominent scientists, an important Yuasa criterion? Or was our country slipping behind?

Table 6.1 shows what I found.

This review of prominent scientific achievements suggests that Yuasa's Phenomenon is not in play again. Only seven years away from the outer edge (2020) of Yuasa's theory, the percentage of American winners in each category was well above the 25 percent level stipulated by Yuasa. Moreover, America's overall portion of 1983–2011 Nobel Prize awards is 48 percent, six points higher than the 42 percent America had earned by 1950.

Youth Rules

This is good news for America, right?

Then I remembered the observation of Guohas and Hongzhou, who emphatically stated, in their review of Yuasa's Phenomenon, "With no exception, the average age of outstanding scientists of any country in the years advancing toward being (or remaining) the world's scientific center does not exceed 50."[12]

Table 6.1 And the Winner Is . . .

Field	Number of Award Winners, 1983–2011	Number of American Winners	American Winners by Percentage
Chemistry	62	26	44%
Medicine	64	31	48%
Physics	71	32	45%
Economics	49	30	61%

Table 6.2 Less Interested

	STEM as a Percentage of All Degrees Conferred		
Year	Bachelors'	Masters'	Doctoral
1966	35.6%	29.2%	64.5%
1970	36.3%	25.7%	61.2%
1980	33.7%	21.6%	57.3%
1990	32.7%	24.3%	63.4%
2000	33.0%	21.3%	62.9%
2004	33.4%	21.6%	62.3%

Under the age criterion, America's prospects for continued global leadership in science and technology immediately became more precarious. Of the 95 living Americans who won Nobel Prizes in chemistry, medicine, physics, and economics in 1983–2011, 11 percent were 50–59 years old when they won, 58 percent were 70–79 years old, and 26 percent were 80 or older. It's a testament to the American education system in the first half of the twentieth century. Only 5 percent of American Nobel Prize winners from 1983 to 2011 were under 50 when they won their prestigious awards—a critical benchmark for continued and aspiring countries to become world scientific centers, according to Yuasa.

Delving deeper in Yuasa's work, I found more ominous news for the continued leadership of America in science and technology. Yuasa's Phenomenon states that there is a direct relationship between a change in the number of scientists and the shifting of the world's scientific center. Yuasa claimed that the world's scientific center always has a concentration of outstanding scientists. Conversely, a scientifically backward country lacks a concentration of prominent scientists.

So how is America's concentration position holding up? As shown in Table 6.2, not well. I did a review of the concentration of the 8,168,623 bachelors', masters', and doctoral degrees conferred in America from 1966 through 2004, comparing all degrees awarded with the degrees awarded in STEM.[13]

In the six years reviewed in the chart spanning a 38-year period, STEM as a percentage of all degrees conferred were down since 1966: 6.2 percent for bachelors' degrees, 26.1 percent for masters' degrees, and 3.5 percent for doctoral degrees.

Look to the East?

If you look at my Nobel Prize analysis, the answer to whether we should look to the East is a resounding *no*. Everybody talks about the rise of China, but from 1983 to 2011, only five Chinese scientists and engineers won Nobel Prizes. And none of them were under 50.

China, which is predicted by many to take over the mantle of scientific leadership from America sometime this century, faces key internal problems, which may extend the 80- to 100-year period outlined by Yuasa. (Note: Germany, the country Yuasa claimed was the world's scientific leader before the United States, had a 110-year reign, from 1810 to 1920.)

Gary Becker, a prominent American economist and senior fellow at Stanford University's Hoover Institute, has this to say about the challenges faced by the People's Republic of China:

> I believe China's continued rapid growth is not assured because of two key concerns: over half of its manufacturing output is produced by state-owned enterprises, plus as Chinese people look at events like the Arab Spring there is cause for concern of social unrest.[14]

I believe there is yet another factor to consider.

Three Patents to the Win

China is not ready to become the world's scientific center for another reason: a weak record in patents. Reuters reported, "China became the world's top patent filer in 2011, surpassing the United States and Japan as it steps up innovation to improve its intellectual property rights track record."[15] It's impressive at first glance, but when you look deeper, it's not.

The Chinese government provides financial incentives to Chinese companies to file patents. Filing a patent with your country's domestic patent office means relatively little in the global market. The big game in patents is what are called *triadic patents*, in which the inventor files for patents to the patent offices of the United States, the European Union, and Japan. The Conference Board in Canada reported in 2007 that 30.7 percent of the world's triadic patents were filed by Americans, followed by Japan at 28.2 percent and Germany at 11.6 percent. The People's Republic of China does not register even a single digit.[16]

The Economist Intelligence Unit reviewed triadic patent filings by how many there were by country per million residents in 1995, 2005, and 2007. The results are shown in Table 6.3.[17]

Although the United States remains the market share leader in triadic patent filings, it falls to seventh place on the global list when analyzed by triadic patent filings per million people. China, though in eighth place, trails the other seven countries by a very wide margin.

Table 6.3   Number of Triadic Patents per Million People

Country	1995	2005	2007
Switzerland	104	107	119
Japan	73	117	114
Germany	59	76	75
Netherlands	48	67	63
Israel	28	60	49
South Korea	7	58	40
United States	45	53	57
China	0.02	0.27	0.30

But another component of Yuasa's Phenomenon suggests a faster possible shift from America to the People's Republic of China: the sheer number of STEM-trained students that China is producing each year.

America's Innovation Ecosystem at Risk

The numbers in Table 6.4 tell a foreboding story for America. In 2004, President George W. Bush directed the President of the United States Council on Science and Technology to study the foundations of America's innovation pipeline. The council expressed significant concern about the total number and concentration of engineering degrees in America compared to other countries, including China:

> Scientific and engineering talent lies at the core of the Nation's innovation ecosystem. Technical skills are required at all points within the

Table 6.4   2004: Engineering Not a Popular Degree in the United States

Country	Total Number of Bachelors' Degrees	Bachelors' Degrees in Engineering	Engineering Degrees by Percentage
China	567,800	219,500	39%
South Korea	209,700	56,500	27%
Taiwan	117,400	26,600	23%
Japan	542,300	104,600	19%
United States	1,253,100	59,500	5%

Table 6.5   Undergraduate Engineering Degrees

China	704,600
South Korea	133,000
Japan	95,200
United States	69,900

China has a tenfold advantage over the United States.

ecosystem from the research labs to the basic workforce. The report emphasizes the strong correlation between mathematics and science education and workforce preparation. Our subcommittee concluded challenges exist at all points in the STEM workforce supply chain, putting our national innovation ecosystem at risk.

The group also reported that "the United States system [of innovation] is threatened by significant changes in the global technical talent pool and the loss of global market share in technical talent."[18]

The United States, by Yuasa's yardstick, while in a tenuous third position, just ahead of South Korea, in the number of engineering bachelors' degrees conferred, is dead last in the important Yuasa criterion of concentration, having only 5 percent of its total bachelors' degrees being in engineering. That's nearly eight times less than the People's Republic of China—not a good omen.

Four years later, in 2008, the National Science Foundation issued a report on the composition of undergraduate degrees in engineering and the natural sciences. As shown in Tables 6.5 and 6.6, it indicated more ominous news for the United States.[19]

Does It Work for You?

A 2005 McKinsey Global Institute report offers some hope to the United States. America's college and university system is recognized as the best in the world. What McKinsey did was to estimate the percentage of engineering undergraduates in the United States and China that were "suitable to work for multinational enterprises."[20]

Table 6.6   Undergraduate Natural Science Degrees

China	297,300
United States	177,600
South Korea	37,700
Japan	35,200

The United States does better in the natural sciences than South Korea and Japan.

Table 6.7   Suitable for Work as Engineers in a Global Firm

Country	Number of Engineering Degrees	McKinsey-Suitable by Percentage	McKinsey-Suitable by Number
United States	69,900	80%	55,920
China	704,600	10%	70,46

Source: Diana Farrell, "Sizing the Emerging Global Labor Market," McKinsey Global Institute, 2005; Science, Technology, Engineering and Math Expansion Program, National Science Foundation, 2008.

Using the 2008 National Science Foundation numbers for undergraduate engineers, I applied the McKinsey numbers to determine the number of "suitable to work for multinational enterprises" students that each country's education system is now producing.

Since this book is about STEM education, I opted to add the number of suitable workforce numbers of engineers and natural scientists to get a big-picture view of the competitive situation the United States finds itself in with China (see Tables 6.7, 6.8, and 6.9).

The United States can take some comfort if the raw degree data is filtered by the McKinsey "suitable to work for multinational enterprises" standard. But what happens if America's education system slips to a 60 percent suitability factor and China improves to 20 percent? What happens if the United States doesn't grow beyond the 80 percent high–watermark?

Not a pleasant thought.

Table 6.8   Suitable for Work as Scientists in a Global Firm

Country	Number of Natural Science Degrees	McKinsey-Suitable by Percentage	McKinsey-Suitable by Number
United States	177,600	80%	142,080
China	297,300	10%	29,730

Source: Diana Farrell, "Sizing the Emerging Global Labor Market," McKinsey Global Institute, 2005; Science, Technology, Engineering and Math Expansion Program, National Science Foundation, 2008.

Table 6.9   McKinsey Math: U.S. Still Leads

Country	Suitable Engineers	Suitable Natural Scientists	Total
United States	55,920	142,080	198,000
China	70,460	29,730	101,190

Table 6.10   Chinese and Indian GDPs Top U.S. GDP by 2050

Country	2009 GDP ($ bn)	2050 GDP	Growth Percentage
China	$8,888	$59,475	669%
India	$3,752	$43,180	1,150%
United States	$14,256	$37,876	265%

The World in 2050

PricewaterhouseCoopers (PwC) provides an economic look at the world in 2050. The report's key conclusion is that the global financial crisis of 2007 has further accelerated the shift in global economic power. One critical measure of a national economy, according to PwC, is something called purchasing power parities, and PwC uses it to estimate future national GDP for countries. Table 6.10 shows a list of the top economies in the world in 2009 and 2050. [21]

Although PwC's numbers for the 2050 GDPs are estimates, if they are accurate, then both the People's Republic of China and India will surpass the economic size of the United States.

America must decide how far down the World Economic Forum list of 142 countries it is willing to fall. We seem to accept the fact that the quality of our math and science teachers is fifty-first in the world. We accept the fact that our students generally test in the middle of the global pack of nations in respected international science and math assessment tests.

Slip Sliding Away?

As we conclude this chapter on Yuasa's Phenomenon, let's compare the United States to the People's Republic of China on the four key components of Yuasa's theory:

1. **A concentration of outstanding scientific and technological achievement.** If looked at through the Nobel Prize lens mentioned earlier, this is a solid advantage for the United States. But comparative research and development trends for areas like clean energy, transportation, and IT seemed to favor China.
2. **A large number of prominent scientists.** With a population five times that of the United States and a higher number of baccalaureate degrees awarded each year in science and engineering, China clearly gets the nod.

3. **Young prominent scientists.** China's huge population advantage also gives it an edge in this category.
4. **A solid education system that rapidly advances young students into scientific careers.** This is a draw, with the United States enjoying the advantages of a more solid education system but being handicapped by stubbornly supporting mediocrity rather than meritocracy in that system.

This very subjective tally puts China ahead in two categories and the United States in one, with a draw in the final category. But if to be a world scientific leader a country needs to lead in all four categories, the shift away from the United States is already happening. This fact also seems confirmed in the 2012–2013 World Economic Forum's *Global Competitiveness Report*. While the People's Republic of China remains far down the list, in just six years the United States has fallen from number one in the World Economic Forum's study to number seven.

Niall Fergusson, a Harvard history professor, has examined what he calls the Four Mores as signposts for China's future. He claims that China's national strategy is based on "consume more, import more, invest more abroad, and innovate more."[22] America cannot match China in the consume, import, and invest abroad categories, but we can and must win in the innovate category.

Survival Is Not Compulsory

Yuasa's Phenomenon will play out again in this century. Perhaps it already is. Will it happen by 2020? No one knows for certain. But remember this component of Yuasa's Phenomenon: While one country may be the overall world leader in science and technology, another country can lead in a different category.

To remain relevant in the twenty-first century and beyond, America needs to utilize its collective mind to out-innovate the rest of the world. We must build a next-generation education system so our country can build the best "empire of the mind," to use the phrase from the Churchill quote that opened this chapter, in the future. We need to become the world's innovation center. According to the World Economic Forum's 2012–2013 *Global Competitiveness Report*, the United States is currently in seventh place on this critical list. We *must* to do better.

Karl Fisch framed it well in his "shift happens" slide show. So did W. Edwards Deming when he said, "Learning is not compulsory and neither is survival." And even though we don't have full data from the SAT because taking it is not compulsory in the United States, other evidence suggests that for nearly five decades, American students have not learned their math and science lessons well, placing our nation's future as a world power in jeopardy.

Notes

1. Winston Churchill, speech at Harvard University, September 6, 1943, Churchill Society, www.churchill-society-london.org.uk.
2. Antoine van Agtmael, "How Can U.S. Stay on Top of the World?," *USA Today*, March 6, 2007.
3. John Gapper, "Capitalism: U.S. No Longer Dominates the World of Business," *Financial Times*, January 23, 2008.
4. Bob Suh, "Opinion: American Companies Are Falling Behind in Technology," *Financial Times*, February 12, 2008.
5. Karl Fisch, "Did You Know: Shift Happens," Fisch Bowl, September 2006. www.fischbowl .blogspot.com.
6. Titus Galama and James Hosek, *Perspectives on U.S. Competitiveness in Science and Technology* (Santa Monica, CA: Rand Corporation, 2007), 1.
7. Jian Guohua and Zhao Hongzhou, "Shifting of World's Scientific Center and Scientists' Social Ages," *Scientometrics*, vol. 8, nos. 1–2, 1985, 59; Yuasa, Mitsutomo. "Center of Scientific Activity: Its Shift from the 16th Century to the 20th Century," *Japanese Studies in the History of Science*, 1962.
8. Ye Feng, Liming Liang, "Shifts in the World Science Centre: Space-Time Characteristics," *Interdisciplinary Science Review*, vol. 25, no. 3, 2000, 1.
9. Guohua and Hongzhou, "Shifting of World's Scientific Center and Scientists' Social Ages."
10. Seymour Papert, Oral History, The Computerworld HonorsProgram, www.cwhonors.org.
11. Guohua and Hongzhou, "Shifting of World's Scientific Center and Scientists' Social Ages."
12. Ibid., 59.
13. American Sociological Association, *Four Decades of STEM Degrees (1966–2004)*, www.asanet .org.
14. Gary Becker, "Is China's Economic Future a Rosy One?" *Becker-Posner Blog*, February 26, 2012, www.becker-posner-blog.com.
15. Lee Chyen Yoe, "China Tops U.S., Japan to Become Top Patent Filer," Reuters, December 21, 2011.
16. "Share of World Patents," Conference Board of Canada, http://www.conferenceboard.ca/ hcp/details/innovation/share-of-world-patents.aspx.
17. Economist Intelligence Unit, "A New Ranking of the World's Most Innovative Countries," *Economist*, April 2009.
18. President of the United States Council on Science and Technology, "Sustaining the Nation's Innovation Ecosystem," White House, June 2004, www.whitehouse.gov.
19. National Science Foundation, Science and Engineering Indicators: 2012 *Digest*, Arlington, VA, January 2012.
20. Diana Farrell, *Sizing the Emerging Global Labor Market* (San Francisco, CA: McKinsey Global Institute, 2005), 8.
21. PricewaterhouseCoopers, *The World in 2050: How Big Will the Major Emerging Market Economics Get and How Can the OECD Compete?*, March 2006, www.pwc.com/gx/en/ world-2050/pdf/world2050emergingeconomies.pdf.
22. Niall Ferguson, "In China's Orbit: After 50 Years of Western Predominance the World Is Tilting to the East," *Wall Street Journal*, November 18, 2010.

CHAPTER 7

1963: SAT Down

The long decline in college board scores has been to education what the Pennsylvania Legionnaire's Disease was to medicine—a mystery that prompts endless speculation and no final answer.

—David G. Savage, "The Long Decline in SAT Scores"[1]

Writing this book has stirred lots of memories for me—like sitting in the overcrowded classrooms of the 1950s, taught by teachers more intent on completing the math or science textbook before the end of the school year than on making sure the students actually understood what was being taught. Another memory is walking into a musty high school cafeteria on a Saturday morning to take the SAT by filling in circles with a number-two pencil. I was a good student in high school, placing in the top 20 in my class of 300 students. But I never tested well.

The History of the SAT

The SAT was introduced into American society by the College Entrance Examination Board in 1926. Twenty-one years later, the Educational Testing Service in New Jersey became the test's official administrator. Although SAT originally stood for Scholastic Aptitude Test, the name was changed to Scholastic Assessment Test in 1994 after an academic debate about the test's nomenclature. The SAT now measures three areas instead of two: math, critical reading (what used to be called *verbal*), and writing, which was added in 2005. Each area is scored on a 200–800 range.

Asleep at the Wheel for 14 Years

In the spring of 1963, six months after President Kennedy's 1962 Rice University lunar landing speech, the first class of baby boomers (born in 1946) filed into cafeterias across America to take the SAT. The group as a whole performed poorly in both the math and verbal portions of the test and this began a 14-year testing disaster as SAT scores continued to decline (see Table 7.1).[2]

The College Entrance Examination Board Responds

This was a turbulent period in American society. The assassinations of President Kennedy, Martin Luther King, and Senator Robert Kennedy; the Vietnam War and the mass protests against it; the civil rights movement and race riots; Watergate; and the Arab oil embargo all occurred in this period and dominated the news media headlines.

The precipitous 14-year drop in the SAT scores did not.

But the College Entrance Examination Board had to pay attention to the results, and in October 1975, Sidney P. Marland Jr., the president of the College Entrance Examination Board, acknowledged the decline when he announced that he was forming a commission to study the 1963–1976

Table 7.1 Fourteen-Year SAT Test Score Decline

Year	SAT Math	SAT Verbal	SAT Total
1963	475	498	973
1964	473	496	969
1965	471	496	967
1966	467	495	962
1967	466	494	960
1968	462	491	953
1969	460	488	948
1970	454	487	941
1971	450	482	932
1972	443	481	924
1973	440	478	918
1974	437	473	910
1975	429	470	899
1976	429	471	900

results. He said, "No topic related to the programs of the College Board has received more public attention in recent years than the unexplained decline in scores earned by students on the SAT test. We are appointing a blue-ribbon panel to assist in making sense out of the complex and inter-related issues involved."[3]

Marland tried to frame the review as independent. But with all 21 of the advisors on the panel personally selected by the College Entrance Examination Board and the Educational Testing Service, it was a tough sell. Few expected an impartial review.

In July 1977, the panel released its 63-page report, *On Further Examination*, with an introduction that said the following: "Starting with the aspects of the score decline, as they seem to permit objective analysis, we try to provide a broader perspective. That means proceeding from what we have been able to establish on the basis of available data to what we believe is a reasonable interpretation of broader evidence."[4] It then referred to the SAT test instrument as the "unchanging standard," which indicated that objectivity had gone out the window.

More Competition for the SAT

The commission was in a difficult position. For more than three decades, the College Entrance Examination Board had had a monopoly in the college entrance test marketplace. However, in 1959, a new test, called American College Testing (ACT), was introduced as a competitor to the SAT. In hindsight, it seems unlikely that the "unchanging standard" SAT was ever set to receive an objective review from the College Entrance Examination Board, because if the commissioners were critical of the SAT and said it was the primary cause of the 14-year decline, ACT would tout that as a reason for more schools to switch to its test. (Note: In 2012 the upstart ACT exam caught up with the SAT, with each group administering 1.6 million tests per year.)

Why the SAT Scores Dropped

The commission identified five reasons for the 1963–1976 SAT scores decline:

1. **Teachers and school systems.** Referring to "what is widely perceived as a serious deterioration of the learning process in America," the commission found that "more and more high school graduates show up in college classrooms, employers' personnel offices, or at

other common checkpoints with barely a speaking acquaintance with the English language and no writing facility at all."[5]

2. **Parents.** The commission criticized American parents who "watch children come home from school without homework to sit passively hour after hour and day after day in front of television sets until they have spent more time there than anyplace else except bed."[6]

3. **Test takers.** The commission concluded that "a major part of the decline in this six-year period [1963–1969] is clearly traceable to a change in the *composition* of the student group taking the test resulting from the deliberate and historic decision in this country in the 1960s to extend and expand educational opportunity and to eliminate previous discrimination in according it."[7] Translation: More racial minorities, women, and low-income students were taking the test, and this resulted in lower test scores for the 1963–1969 test period.

   There is only one problem with this explanation: it makes no sense for that period. The commission was referencing the historic bill known as the Elementary and Secondary Education Act, which opened the educational opportunity doors to millions of young Americans. It was signed into law in April 1965, and the portion of graduating high school seniors opting to take the SAT test surged from 52 percent in 1965 to 68 percent in 1966. So the compositional argument offered by the commission would not have made sense until 1966. But the rationale then evaporated in 1967, the year I took the SAT, because the portion of graduating seniors taking the test fell back 13 percentage points to 55 percent (still a healthy 8 percent increase from the 1964 figure of 47 percent).

4. **American society.** The College Entrance Examination Board claimed that "pervasive" factors in American society were responsible for lower SAT scores. The commission staked out this wishy-washy argument as follows: "Although the test scores do not themselves indicate the nature of these [pervasive] forces, there are available data that permit informed conjecture."[8]

   So just what exactly were the "pervasive" factors identified by the commission? It described them as "changes that have taken place in schools and society during a period of turbulence and distraction rarely paralleled in American history."[9] This clearly refers to the events I mentioned earlier in the chapter.

5. **Media.** The commission also assigned blame to "the advent of television," because "by age 16 American children have spent between 10,000 and 15,000 hours watching television, more time than they have spent in schools."[10]

Parents, teachers and school systems, lower standards in education, a turbulent American society, and lazy kids who spend countless hours passively in front of television sets—all were to blame, but not the "unchanging standard," the sacrosanct SAT itself. Not much happened after the report was released. The relative importance of the report was perhaps expressed best by the commission chairman and former secretary of labor Willard Wirtz, who summarized the findings in *Time* magazine by saying that America was "off stride for 10 years." Well, actually it was 14 years, Mr. Secretary. But who's counting?

How to Get 100 More SAT Points

SAT scores stagnated, rose slightly, and then fell during the 1980s and the early 1990s. Confused and bewildered by the weak results, the College Entrance Examination Board came up with a solution: it "recentered" all SAT scores before 1995. I can just hear the readers saying, "What is recentering?" That's the same question I had when I first read about it. Let me explain.

Both components of the SAT before 1995, math and verbal, were scored from 200 to 800, with the median score being 500. By 1995, however, the median scores had dropped to 428 in verbal and 478 in math. Enter recentering, a statistical process whereby the stat wonks at the College Entrance Examination Board added 72 points to all pre-1995 verbal scores and 22 points to all math scores in order to bring the median scores back to 500. (Note: Go to your favorite search engine and type in "SAT recentering" to find out more about this. If you took the test before 1995, you can add 100 or more points to your SAT score!)

Too Much Mediocrity

Although readers who took the test before 1995 might be happy with a recentered SAT score, others were not. *Newsweek* columnist Robert J. Samuelson slammed the recentering decision with these sharp words: "We have plenty of random mediocrity in America; we don't need to promote it. . . . Almost everyone gets a boost. . . . This is nationalized grade inflation" that sanctions mediocrity, "blurs the distinction between students' abilities in math and reading," and "will obscure long-term trends and make them harder to explain."[11]

In 2011 the average SAT math score was 514, or equivalent to a precentered score of 492. Verbal was worse, with the average score being 498, a precentered score of 426, the lowest since the College Entrance Examination Board started the SAT test in 1926.

On Further Examination addressed five reasons that American high school students performed miserably in the SAT test for 14 years, starting in 1963, and these reasons have logical roots. But the explanations are dwarfed by the overarching reason that the American school system continues to languish: no clear lines of accountability exist for who is ultimately responsible for improving public education in America. This challenge will be addressed in the next chapter.

Notes

1. David Savage, "The Long Decline in SAT Scores," *Educational Leadership*, January 1978.
2. College Entrance Examination Board, *On Further Examination: The Report of the Advisory Panel on the Scholastic Aptitude Test Score Decline* (New York: College Entrance Examination Board, 1977), 13.
3. Ibid., 4.
4. Ibid., 8.
5. Ibid., 19.
6. Ibid., 21.
7. Ibid., 24.
8. Ibid., 25.
9. Ibid., 28.
10. Ibid., 31.
11. Robert J. Samuelson, "Merchants of Mediocrity: The College Board Nationalizes Grade Inflation," *Newsweek*, August 1, 1994.

CHAPTER 8

1976: Too Many Chiefs

Education is a local responsibility, a state function, and a national concern.

—NORMAN C. THOMAS, *EDUCATION IN NATIONAL POLITICS*[1]

When I meet with IT executives, I enjoy asking them (I suppose *teasing* them is a more accurate description), "When was the U.S. Department of Education founded?"

Most haven't a clue. Those who venture a guess usually place it in the New Deal era under President Roosevelt or in the Kennedy administration.

Here are the facts. From 1908 through 1975, 13 American presidents proposed 130 bills to Congress to create a U.S. cabinet-level department of education. All of them failed. Then a political quid pro quo succeeded in creating the U.S. Department of Education in November 1979.

Most people are shocked to learn this.

A Tale of Two Documents

I enjoy visiting the northwest region of the United States, particularly in the summer months, when you can stand in downtown Seattle and it seems as though you can reach out and touch the top of Mount Rainier 50 miles to the southeast. In 1787, however, the northwest portion of the United States of America consisted of Ohio, Illinois, Indiana, Michigan, Wisconsin, and parts of Minnesota.

As this newly formed region of the United States was being settled, a framework of law known as the Northwest Ordinance of 1787 was created on July 13, 1787. A key provision of the Northwest Ordinance was that "religion, morality and knowledge being necessary to good government and the happiness of mankind, schools and the means of education shall be forever encouraged."[2]

Just 66 days later, on September 17, 1787, another document was signed into law: the Constitution of the United States. And on one key issue, the role and responsibility for public education, the Northwest Ordinance of 1787 and the U.S. Constitution were at odds. In the 4,543 words of the Constitution, there is no direct mention of the words *education* or *educate*.

In an essay published just days after President Lyndon Johnson signed the 1965 Elementary and Secondary Education Act, *Time* magazine commented on the different ways in which the two laws looked at education: "The United States Constitution, drafted the same year as the Northwest Ordinance, said nothing about education, reserving that function to the states—which assumed the task so conscientiously that, even without federal direction, the uniquely American drive toward universal education soon became a key strength of the nation."[3]

Keep It Local

With the line drawn in the sand, state and local municipalities took the lead in educating Americans. At the federal level, there was the U.S. Office of Education, created as a statistics clearinghouse in the Department of the Interior (1868); the Federal Security Agency (1939); and the Department of Health, Education, and Welfare (1953), which included an educational analysis unit with no broad education policy-making powers. The U.S. Office of Education gathered data and made trade mission trips like the trips to the Soviet Union in the late 1950s to learn how that country taught math and science to its young people.

The Great Society Era Ushers in Federal Involvement

President Johnson's Great Society program included the Elementary and Secondary Education Act (ESEA), which put a keener focus on the importance of education at the federal level. At the bill's signing ceremony in April 1965, Johnson underscored the significance of the bill when he said, "I will never do anything in my entire life that excites me more or benefits the nation I serve more, or makes the land and all of its people better and wiser and stronger or anything that I think means more to freedom and justice in the world than what we have done with this education bill."[4]

ESEA: Not All Things Considered

Just what was the Elementary and Secondary Education Act? It wasn't really an education act per se. It was mostly a national primary and secondary school funding act set to be reauthorized on a regular basis by Congress. An important element of the act was the explicit hands-off posture of the federal government. The new law explicitly forbade the establishment of a national curriculum, a debate that is raging on in 2013 American education politics.

The ESEA was created to distribute $1.3 billion to schools and school districts across America that met the qualification that 40 percent or more of the student population comes from low-income families.[5] The bill aimed to provide underprivileged Americans with an education opportunity. And in an amazing example of how politicians in Washington can manipulate an original piece of legislation, by 2012, any school district in America with at least 2 percent of its students in poverty received Title I Basic funds.[6] Even the most affluent school districts in the country can now qualify for Title I funds.

It must be emphasized that even though ESEA raised the profile of education as a key national issue, it was largely a congressional *funding* bill. In its original version, ESEA did not give broad education powers to the president of the United States.

Teacher Unions Create the U.S. Department of Education

By the mid-1970s, the major teacher unions, the National Education Association and the American Federation of Teachers, witnessed a significant growth in membership largely as a result of the power of collective bargaining. Although the teacher unions were focused on state and local political issues, the national leaders of the unions, with millions of members under their control and hundreds of millions of dollars in collectible annual dues, decided to test the national political waters.

The National Education Association challenged, and successfully blocked, President Richard Nixon's Supreme Court appointees, G. Harrold Carswell and Clement Haynsworth, in 1970. Two years later, however, the National Education Association and the American Federation of Teachers sat out the 1972 Nixon versus McGovern presidential campaign. Perhaps they didn't want to waste political capital supporting Senator George McGovern, who was considered a weak candidate.

The situation changed in 1976, when the National Education Association threw its full national, state, and local political clout behind helping Georgia governor Jimmy Carter get elected president. What the teacher unions wanted in return for their support is very clear from the title of a 1975 report coauthored by the National Education Association and a group it helped create called the Labor Coalition Clearinghouse: *Needed: A Cabinet Department of Education*.

Did I Really Promise *That?*

Many candidates make promises in the heat of a campaign only to forget the promises after the election. And so it was with President Carter, who campaigned on a platform of smaller government. The newly elected president was not interested in expanding the federal government with the creation of a new cabinet-level education department. Nor was the proposed new education department endorsed by Joseph Califano, Carter's secretary of health, education, and welfare, who said in an interview, "My concern is it threatens to breach the healthy limits of federal involvement in education. I also think the National Education Association and the Labor Clearinghouse Coalition gave insufficient weight to the wisdom of keeping primary responsibility for elementary and secondary education on state and local community levels."[7]

In Washington, reports like *Needed* are a dime a dozen. What counts is votes, and in the 1978 midterm elections, President Carter and the Democratic Party took a beating, losing 15 House seats and 3 Senate seats to the Republicans. Carter, though not a masterful politician, was smart enough to understand this reality: if he wanted to be reelected in 1980, he needed all the help he could get. Proposals championed by Vice President Walter Mondale to create a U.S. Department of Education had failed to get congressional support in the first three years of Carter's presidency, but after the disastrous midterm results, relentless lobbying by the National Education Association led to the creation of a separate department, and on October 17, 1979, the bill to create the first U.S. Department of Education passed in Congress.

The rationale to create the department was purely political. Shortly after the bill was passed, the *Wall Street Journal* interviewed a Democratic member of Congress who said, "The idea of an Education Department is really a bad one. But it's the National Education Association's top priority. There are schoolteachers in every congressional district [more than 4,000 at the time], and most of us simply don't need the aggravation of taking them on."[8]

President Carter's Top 10 List

Here are the 10 reasons the Carter administration believed that a cabinet-level Department of Education was necessary, according to the bill that created it:

1. Education is fundamental to the development of individual citizens and the progress of the Nation.

2. There is a continuing need to ensure access for all Americans to educational opportunities of a high quality, and such educational opportunities should not be denied because of race, color, national origin, or sex.

3. Parents have the primary responsibility for the education of their children, and States, localities, and private institutions have the primary responsibility for supporting that parental role.

4. In our Federal system the primary public responsibility for education is reserved specifically to the States and the local school systems and other instrumentalities of the States.

5. The American people benefit from a diversity of educational settings, including public and private schools.

6. The importance of education is increasing as new technologies and alternative approaches to traditional education are considered as society becomes more complex.

7. There is a need for improvement in the management and coordination of Federal education programs to support more effectively State, local, and private institutions.

8. The dispersion of education programs across a large number of Federal agencies has led to fragmented, duplicative and often inconsistent Federal policies relating to education.

9. Presidential and public consideration of issues relating to Federal education programs is hindered by the present organizational position of education programs in the executive branch of the government.

10. There is no single, full-time Federal education official directly accountable to the President, the Congress, and the people.[9]

Read that list again. Can you find one reason that articulates how the creation of the Department of Education would improve public, or private, education in the United States? The fourth point is particularly profound; it clearly indicates that the primary responsibility for education lies with the states and local school districts, not the federal government.

The Department of Education was created as a political favor to the National Education Association. And it was passed by members of Congress who were too wimpy to oppose an average of 4,000 teachers per district.

Eight Years Is Too Short

I now want to focus on the reason offered in the tenth point, the need for a single federal education official. The biggest problem I have with the

Department of Education is that it is controlled by the political agenda of the president, not the interests of the 49 million schoolchildren in America.

With the president guiding the nation's education strategy, America is beholden to an educational policy that defines *long-term* as eight years—because that's the longest possible presidential term.

Since the Department of Education was created in November 1979, the United States has had six education strategies. Let's take a look at some of them.

Reagan Shifts from Compliance to Competency

President Ronald Reagan tried, unsuccessfully, to dismantle the Department of Education in his first term. But then his administration did something in his second term that continues to affect American education in the twenty-first century. The reauthorization of ESEA came before Congress. With William Bennett, the third secretary of education, leading the way, Congress shifted the prime focus of ESEA from compliance with federal law, as the prime determinant of receiving ESEA Title I funds, to competency.

Welcome to the dawn of the "teaching to the test" era.

Bush Sets Voluntary Education Goals

President George H. W. Bush supported Secretary Bennett's and President Reagan's policies in putting a significant national focus on student and teacher competency. In January 1989, immediately after his inauguration, President Bush hosted an education summit attended by all 50 governors. Ray Scheppach was the executive director of the National Governors Association, and as the summit opened, he framed the goal of the meeting as "establishing consensus around a set of national goals for education improvement, stated in terms of the results and outcomes we as a nation need for the education system."[10]

Thus, in just one decade, the federal government went from being a dispenser of Title I funds authorized by Congress for low-income areas to the proposition that it be the nation's arbiter of national education goals.

Because so much of the focus on "teaching to the test" began in this era, it is worthwhile to examine the six goals of the 1989 education summit:

1. Ensure that all children start school ready to learn.
2. Achieve a high school completion rate of 90 percent (which has been achieved nationally).

3. Improve achievement for all Americans in all basic subjects.
4. Make American students first in the world in math and science (far from being achieved).
5. Ensure that all adults are literate and have access to lifelong learning.
6. Make schools safe, disciplined, and drug-free.[11]

Other Issues Get in the Way

Lamar Alexander, President Bush's secretary of education, framed the goals of the summit into a piece of legislation called America 2000. But what happens when an issue like education is on the president's plate and other issues arise that demand immediate attention? Education, which has never been a critical national priority, is often set on the back burner.

For example, 18 months after Bush's education summit, Iraq invaded Kuwait on August 2, 1990. A massive coalition effort, led personally by President Bush, became the administration's top priority and, in fact, defined and consumed the remainder of Bush's one-term presidency. America 2000 quietly died in the House and the Senate.

Clinton Unsuccessfully Shifts Education Goals from Voluntary to Compulsory

Under President William J. Clinton, America 2000 morphed into Goals 2000. Goals 2000 aimed to promote the achievement of national education goals by the turn of the twenty-first century. Goals 2000 was different from America 2000 in one significant component: whereas America 2000 proposed the voluntary participation of the states in national achievement tests, Goals 2000 sought to make that participation compulsory. That's a huge difference.

It presented education secretary Richard Riley with a monumental challenge: How does the federal government, with no rights regarding education mentioned in the Constitution, tell the states what to measure and how to measure it? Just as with President Bush, however, other issues took President Clinton's attention away from national education: the sweeping political defeat in the 1994 midterm elections, the domestic terrorist attack in Oklahoma City, and the Whitewater and Monica Lewinsky investigations, to name just a few.

I had the opportunity to meet several times with Secretary Riley as he pushed hard to make the technology component of the 1994 Improving America's Schools Act a reality. He is an outstanding and capable person. But with all the other issues consuming Washington, Riley shifted the focus of

Goals 2000 in President Clinton's second term to tamer issues like charter schools.

No Child Left Behind Ushers in Compulsory Education Compliance

On his second day in office, President George W. Bush brought the issue of education back into the national spotlight. It had been seven years since ESEA had been reauthorized. Bush made a bold move that was intended to show the country he meant his campaign slogan that he would be the nation's first "education president."

On January 23, 2001, President Bush sent the No Child Left Behind bill to Congress as his proposed reauthorization of ESEA. Within a year and with strong bipartisan support, the bill was signed into law. Its most controversial component was an extension of President Clinton's hoped for, but never realized, desire to make a national achievement test in reading, math, and science compulsory.

Moreover, whereas the Clinton administration had gotten bogged down in what measurements to use, the Bush administration, extending a longtime Department of Education testing program called the National Assessment of Educational Progress (NAEP), laid down in clear but controversial terms what the states were required to do to assess all students in grades three through eight in the three subjects. No Child Left Behind categorized achievement as advanced (A+), proficient (A), basic (B), and below basic (D or lower).

The act then set a very high bar: to continue to receive Title I federal education funds, a state was required to bring *all* of its students to proficient level in the three subjects by 2014. As of 2014, if an individual school failed to reach this level for two years in a row, the parents and students would be offered the choice of another school to attend.

Obama Is Stymied by Gridlocked Washington

Enter Barack Obama as president, who was forced from day one of his presidency to focus on the most serious economic crisis since the Great Depression. In early 2010, however, President Obama sent to Congress a major overhaul of the No Child Left Behind Act with this key provision: a call to the states to adopt new academic standards that focused on making sure all high school students are ready for either college or a job by the time they graduate.

Many states applauded this shift as a much more realistic national education goal than the previous standard of 100 percent proficiency in reading, math, and science. But in a politically gridlocked Washington, neither the House nor the Senate took any action on reauthorization of the act, leaving the White House to promote smaller, though more focused, education initiatives like Race to the Top and the massive American Recovery and Reinvestment Act.

Gridlocked Washington may have unintentionally empowered the role of the president in setting administration-based national education goals. Since most states realized that 100 percent proficiency was unattainable, 36 of them approached the Department of Education and asked for exemptions. This led to negotiations between state and federal education leaders. The *New York Times* suggested that the White House won those negotiations.[12] Thus, the federal government gained even more influence over national education policy.

Close Down the U.S. Department of Education

In concluding this chapter, may I offer several personal observations? First, nowhere in the Constitution does it stipulate that education is a federal government responsibility. Second, the Department of Education exists only because of a political favor that Carter did for the National Education Association. Third, and perhaps most important, our country needs a national education strategy and policy that transcends an eight-year period, the maximum number of years an American president can serve.

The United States spends $583 billion on K–12 education each year. Only 8.5 percent, or $49.6 billion, is funding from the federal government. The remainder of the funds come from local ($250 billion) and state ($284 billion) tax coffers. The Department of Education needs to go back to its former home as the nation's keeper of all sorts of education statistics that state and local educators can leverage to improve public education in their towns and states.

The creation of the U.S. Department of Education as a cabinet-level agency just added more confusion about who is ultimately responsible for public education in America. This atmosphere of confusion allowed 33 years of warnings to be ignored that something is broken in how America educates its schoolchildren. You are about to review this litany of warnings.

Notes

1. Norman C. Thomas, *Education in National Politics* (Philadelphia: McKay, 1975), 19.
2. "The Northwest Ordinance," Early America, www.earlyamerica.com.

3. "The Big Federal Move into Education," *Time*, April 30, 1965.
4. Ibid.
5. U.S. Department of Education, "The Elementary and Secondary Education Act," April 1965, www2.ed.gov/policy/elsec/leg/esea02/index.html.
6. U.S. Department of Education, "ESEA Title 1 LEA Allocations," 2012, www2.ed.gov/about/overview/budget/titlei/fy12/index.html.
7. Daniel E. Ponder, *Good Advice: Information and Policy Making in the White House* (College Station, TX: Texas A&M University Press, 2000), 76.
8. David Boaz, *Cato Handbook on Policy* (Washington, DC: Cato Institute, 2005), 284.
9. Department of Education Organization Act 96-88, vol. 210 Cong. Rec., S.668 (daily ed., Oct. 17, 1979).
10. Maris Vinovskis, *The Road to Charlottesville: The 1989 Education Summit* (Darby, PA: Diane Publishing, 1999), 28.
11. Department of Education Organization Act 96-88.
12. Motoko Rich, "'No Child' Law Whittled Down by White House," *New York Times*, July 6, 2012.

PART TWO

And the Hits Just Keep on Coming

As a child, the Aesop fable *The Boy Who Cried Wolf* was one of my favorite stories. You remember the tale. It was about a shepherd boy who repeatedly tricked his neighbors into believing that a wolf was attacking his flock, so when a wolf really did attack his flock, they didn't believe his cry for help anymore. While the moral of the story that persistent liars will not be rewarded does not concern the focus of this book, the call of repeated attacks does, with one key difference: the warnings in this book are real, not imagined.

I started to write this book after reading one such call in 2007, an industry report that questioned America's future competitiveness. Over the next five years I discovered 61 other reports, many with ominous titles referencing storms, cliffs, and hurricanes and written by many smart people. The report conclusions, however, seemed to be written by the same person: a singular voice was saying loud and clear to anyone who cared to listen that over the past half century, America has become complacent and is rapidly losing its position as the world's leader in technology innovation and business competitiveness.

Many of the reports blamed our nation's slow but steady decline on the K–12 public education system in the United States. This education system, the critics said, was more suited to the nineteenth century than the twenty-first. My research also introduced me led me to 18 domestic and international math and science assessment tests spanning the period from 1964 to 2012. The test results confirmed the suspicions of many twenty-first-century American business leaders: American students were not proficient in math and science and were falling further and further behind students from industrialized and emerging countries.

Can You Hear Me Now?

The collective mass of those reports and test results—from respected business leaders, credible testing agencies, governmental agencies, trade associations, and public opinion groups—had a profound effect on me. I kept asking myself, "Is anyone listening? Why does it seem that nothing is being done to address such an obvious problem? Does anyone in America *really* care?"

Road Trip

You are about to go on a 49-year journey, separated into four chapters, through these prominent reports and tests. I present the results of the reports and test scores in chronological order for a key reason: as you review the material, I want you to pay close attention to three issues. First, look at the report titles; you will discover that they become more ominous year after year. Second, take note of the report and test frequency, particularly in the past eight years. The drumbeat warning America about its failures in science and math education keeps getting louder and more frequent with each passing year. Third, notice how little improvement there has been in math and science test scores in nearly half a century. It's rather remarkable when you consider that the United States spends $583 billion on public education each year.

The Eighth-Grade Focus

You will notice that most of the assessment test results focus on eighth-grade students. There are two reasons for that. First, after the eighth-grade assessment, the next set of test results is for twelfth grade, and evaluating American twelfth graders in relation to twelfth graders from other countries is not an apples-to-apples comparison, because other students are often older than American high school seniors. Second, middle school math and science teachers hold the key to America's economic, innovation, and national security future.

U.S. student scores consistently fall from fourth grade to eighth grade, just when the opposite should be happening: in middle school, as the average American student is placed with specific teachers of math and science, one would surmise that these subject-specific teachers would be able to build on the knowledge of elementary school students. Unfortunately, the opposite is the norm: As the average American student progresses through our nation's public school system, he or she performs less well on both domestic and international math and science assessment tests. America desperately needs to reverse that trend.

Connect the Dots

As I reviewed the reports and test results, a thought occurred to me: too often a report or a test result is released to the public and debated for a relatively short amount of time and then forgotten. I decided to take a different approach. I intend to connect these data points into a bigger narrative—a four-part narrative that claims the following: (1) the American education system has done a poor job teaching math and science to middle school and high school students for nearly five decades, (2) those inadequacies are now manifesting themselves on respected international math and science assessment tests, which show American students falling further behind the rest of the world, and on domestic science and math tests sponsored by the U.S. Department of Education, (3) business leaders are losing patience with the ability of the U.S. education system to systemically restructure education to teach the skills needed in a twenty-first-century global marketplace, and (4) what is at stake are the future economic, workforce, and national security issues facing America.

It Takes a Village That Cares

The villagers in *The Boy Who Cried Wolf* warned the boy to "save your frightened song for when there is really something wrong! Don't cry 'wolf' when there is *no* wolf!"[1] A solitary wolf does appear at the end of the story to scatter the flock of sheep; however, in our story *two* wolves have roamed unchallenged in America's pasture since 1964: a failing American education system and increased global competition.

In the fable there is one shepherd boy warning the country; in our story there are scores of them: the authors of the warnings you are about to read. And while the villagers in the fable promptly responded to the perceived threats to the flock, that is unfortunately not the situation in twenty-first-century America, because politicians, businesspeople, parents, teachers, and school administrators have ignored the warnings.

The Warning System Works

Norman Cousins, a twentieth-century political writer, once said, "History is a vast warning system."[2] The data in the next two chapters prove conclusively that America's warning system functions well.

Whether anyone is paying attention to it, however, is another matter.

Notes

1. "The Boy Who Cried Wolf," *Aesop's Fables* (London: Routledge & Sons, 1867), 42.
2. Norman Cousins, *Saturday Review of Literature*, April 15, 1978.

CHAPTER 9

The Skills Gap Warnings Begin

I hold no copyright on the notion that the situation in public education is not a healthy one. Indeed, if one were to select the most shopworn question in American life today, "What's wrong with our schools?" would undoubtedly be a strong contender for top honors.

—MYRON LIEBERMAN, *THE FUTURE OF PUBLIC EDUCATION*[1]

1964: The First International Mathematics Study

Created just as the baby boom generation was coming to a close, the First International Mathematics Study was, as its name says, the initial international mathematics assessment test. Conducted in 12 countries by the International Association for the Evaluation of Educational Achievement, it reported on the math skills of eighth graders and high school seniors with the goal of identifying the factors within each country that led to math proficiency.[2]

In America, a country struggling at that time with political assassinations, race riots, civil rights marches, and an escalating war in Vietnam, the results of the First International Mathematics Study barely caught anyone's attention—even though American eighth graders came in next to last in the study (see Table 9.1).

1971: The First International Science Study

The First International Science Study, also conducted by the International Association for the Evaluation of Educational Achievement, was part of a

Table 9.1   We're Number 1—Plus 10!

Country	Mean Number of 70 Items Answered Correctly
Israel	32.3
Japan	32.2
Belgium	30.4
Finland	26.4
Federal Republic of Germany	25.4
England	23.8
Scotland	22.3
Netherlands	21.4
France	21.0
Australia	18.9
United States	**17.8**
Sweden	15.3

Source: National Center for Education Statistics, *International Mathematics and Science Assessment: What Have We Learned* (Washington, DC: U.S. Department of Education,1992), 12.

larger research project known as the Six Subject Survey, which measured skills in verbal ability, reading comprehension, reading speed, and word knowledge.[3]

The results were mixed for America. In "core science items," U.S. eighth graders were number 7 out of 14 countries participating in the study. American high school seniors did much worse. They came in last!

1971: The National Education Trust Fund

Milton Shapp, a politician who was credited with giving President Kennedy the idea that eventually became the Peace Corps, had another big idea. In 1971, Shapp, the Democratic governor of Pennsylvania, proposed to President Nixon the creation of the National Education Trust Fund for the purpose of improving public education in the United States. Shapp modeled his idea after the Highway Trust Fund, which was created in 1956 to underwrite the construction and maintenance of America's interstate highway system. Edward Simon, Governor Shapp's special assistant for fiscal and economic activity, formally presented the idea to Nixon's staff:

We propose the creation of a National Education Trust Fund as the vehicle for the massive investment in education that is required. The Fund would finance a portion of the costs of education at all levels, and those who benefited most would replenish the Fund through a tax on their incomes throughout their working years. . . . The National Education Trust Fund would advance money only for the direct costs of education and not for such purposes as school construction; the aim of the Fund is to invest in people, not buildings.[4]

I am so intrigued by the governor's idea that I plan to present it in the Epilogue of this book. Forty-two years later, it is the right idea for the right time.

1978: The Nation's Report Card

The U.S. Department of Education has conducted periodic assessment tests in math and science among students in fourth grade, eighth grade, and twelfth grade since 1969. The department has labeled these tests the Nation's Report Card.

Four Performance Levels

In the 1978 assessment, there were four mathematics performance levels for American eighth-grade students. Each level was matched to a range of quantitative scores, as reported by the National Center for Education Statistics:

1. **Ability to do moderately complex procedures and reasoning (score of 300 or higher).** "Students at this level are developing an understanding of number systems. They can compute with decimals, simple fractions, and commonly encountered percents."
2. **Ability to do numerical operations and begin problem solving (250–299).** "Students at this level have an initial understanding of the four basic operations (addition, subtraction, multiplication, and division)."
3. **Ability to develop beginning skills and understanding (200–249).** "Students at this level have considerable understanding of two-digit numbers. They can add two-digit numbers but are still developing an ability to regroup in subtraction."
4. **Ability to do simple arithmetic (under 200).** "Students at this level know some basic addition and subtraction."[5]

Table 9.2 Less than 20 Percent Can Compute with Simple Fractions

Category	1978	1982	1986
Can do moderately complex procedures	18%	17%	16%
Can do numerical operations	47%	54%	57%
Can develop beginning skills	30%	27%	26%
Can do simple arithmetic	5%	2%	1%

Little Ability

Of these four ability categories, even the most skilled category, the ability to do moderately complex procedures, seems rather rudimentary. Yet the abilities of American students are abysmally low in some cases. Table 9.2 shows the math scores, our nation's report card, for American eighth graders in 1978–1986.[6]

1982: The Second International Mathematics Study

The Second International Mathematics Study "provoked considerable controversy when it revealed American students were distinctly mediocre in mathematics compared to peers in most other countries."[7] American eighth-grade students were number 12 out of 14 regions in the study.

The International Association for the Evaluation of Educational Achievement had waited 18 years before conducting its second global mathematics assessment test, in 1982. Fourteen countries or provinces participated in the study. When Japan, a country whose economic influence surged in the 1960s and 1970s, came out on top, alarm bells went off in America.

The study's summary table reveals the number of questions answered correctly by eighth graders from the 14 regions (see Table 9.3).[8]

Table 9.3 Bringing Up the Rear—Again

Country	Average Percentage of Questions Answered Correctly by Eighth Graders
Japan	64%
Netherlands	58%
France	57%
Belgium (Flemish)	56%
Hungary	56%

Country	Average Percentage of Questions Answered Correctly by Eighth Graders
British Columbia, Canada	55%
Belgium (French)	54%
Average of the 14 regions	52%
Ontario, Canada	51%
Scotland	49%
England and Wales	49%
Finland	49%
United States	48%
New Zealand	45%
Sweden	42%

The Japanese were the clear winners of the Second International Mathematics Study. Their strong showing, and the lackluster U.S. performance, caused a stir in America. One commentator wrote, "There is a strong element of competition because Japan has won every section. This idea of winning and losing is a boon to politicians and the media. Words like 'falling standards' are bandied about and teachers are yet again castigated."[9]

1983: *A Nation at Risk*

If an unfriendly foreign power had attempted to impose on America the mediocre educational performance that exists today, we might well have viewed it as an act of war.

—GERALD HOLTON, *A NATION AT RISK*[10]

The seminal critique of American education is a document entitled *A Nation at Risk: The Imperative for Educational Reform*, released by the U.S. Department of Education in April 1983. Among IT executives, business leaders, and the population at large, it remains an unknown report.

A Nation at Risk was written in 1982–1983 by a special commission appointed by Secretary of Education Terrel H. Bell. To appreciate the report's full significance, the reader needs a bit of context. Bell was appointed as Reagan's second secretary of education and tasked with one assignment by Edwin Meese, the counselor to the president (and later attorney general): "Some day early in the administration, walk into the Oval Office and hand the president the keys to the Department of Education and say, 'Well, we've shut the abominable thing down. Here's one useless government agency out of the way.'"[11]

Deming's Unintended Consequence

Although Bell shared President Reagan's misgivings about the relevance of the U.S. Department of Education, he wasn't totally convinced that the agency should be disbanded. But if anyone was going to do it, he was, so he accepted the position. With the Great Recession of 2007 still fresh in our minds, it is easy to forget the economic woes the country faced in January 1981 when Ronald Reagan was inaugurated. The inflation rate was 11.8 percent, and the unemployment rate was 7.6 percent. These statistics, when combined, were often referred to as America's "misery index."

Moreover, the Japanese were continuing to eviscerate the American automobile, office automation, and steel industries, in large part because of that country's flawless execution of W. Edwards Deming's total quality management practices, which Japanese business leaders had embraced 31 years earlier.

These management practices were still being ignored by American business leaders in 1981.

America Wasn't Better Off

The Department of Education had 10 regional offices, and Secretary Bell traveled to each office. He wanted to see firsthand why Americans elected Reagan, who had campaigned with the provocative question "Are you better off now than you were four years ago?"[12] It was a career-changing tour for Bell, as he describes in his memoirs:

> It was in the context of this pervasive decline, and the widespread questions it raised about what was wrong with our government, our people, our labor force, our businesses, and our schools, that I was hearing constant complaints about education and its effectiveness. Our loss of zest and drive and spirit would not be regained until we renewed and reformed our schools.[13]

Time for Action

Bell had heard enough. It was time to act. But he didn't want to follow the tried-and-true, kick-the-can-down-the-road approach to problems: "When in doubt, do a survey of the problem to buy time." Bell wanted something bigger and better. "I wanted to stage an event that would jar the people into action on behalf of their educational system. Since I could not realistically plan on another *Sputnik*-type occurrence, I had to search for an alternative."[14] Bell's first solution—a presidentially appointed commission—was rejected out of hand by James A. Baker, Reagan's chief of staff (and later secretary of the treasury), and Meese because, they said, the president believed that education was a local and state responsibility.

Bell had another approach. As secretary of education, he could appoint a cabinet-level commission to study the problem without asking Meese and Baker for permission. And that's what he did. Bell was aware that a cabinet-level commission risked issuing findings that might be lost in the hundreds of reports that emanate daily out of Washington. He was convinced, however, that regardless of where the commission's authorization came from, the problem it addressed would capture the attention of the nation.

He wanted a high-profile chairman for the commission, so he selected David Pierpont Gardner, the president of the University of California, Berkeley, to lead the effort. Gardner also wanted big names on the commission, so he filled the posts with William O. Baker, the past president of Bell Laboratories; Bartlett Giamatti, the president of Yale University; Gerald Holton, a physics professor at Harvard University, who wrote the final report; and an extraordinary array of talented people that included high school teachers and principals from all regions and levels in the education system.

It was definitely not like the one-sided commission that had been appointed to review the SAT situation several years earlier. Bell wanted a frank and honest appraisal of America and the country's education system. Privately, he was hoping for a report that would be, as he put it, a "*Sputnik*-type occurrence." But he was careful not to convey his goals to Gardner and the other members of the commission.

After an unimpressive first meeting with Reagan at which the president instructed the commission "that we should bring prayer back into the school room and we should attend to the need for vouchers for tax credits," Gardner and his team held hearings throughout the United States, as Bell had done, gathering opinions from thousands of people.[15]

Moreover, the commission itself met face-to-face eight times, sometimes contentiously, before releasing its report, *A Nation at Risk*, on April 26, 1983, in a brief ceremony in the Oval Office.

The Report Strikes a National Nerve

A Nation at Risk immediately struck a national nerve when it hit the wire services and prominent national newspapers. Headlines like "Education Panel Sees 'Rising Tide of Mediocrity'" (Associated Press), "U.S. Education: Unsatisfactory" (*Miami Herald*), "Failure in Education" (*Washington Post*), and "Teaching Teachers: Many Think It's Key to Better Schools" (*Lexington Herald-Leader*) appeared immediately on front pages and in editorial columns across the country.[16]

Not all of the reviews were positive.

Conservative columnist William F. Buckley criticized the report for offering solutions that "you and I would come up with over the phone." Russell Baker, the popular *New York Times* humor columnist, contended that a sentence containing a phrase like "a rising tide of mediocrity" wouldn't be worth "more than a C in tenth-grade English."[17]

But tenth graders can sometimes surprise you with their writing skills. And so powerful was the relatively short report, purposely written in bellicose terms by Gerald Holton, that it got the attention of President Reagan, who seemed to like bellicose language. Only one month earlier, Reagan had told the National Association of Evangelicals in Orlando, Florida, to "beware of ignoring the facts of history and the aggressive impulses of an evil empire," referring to the Soviet Union.[18]

The report completely changed the education plank in the 1984 Republican Party platform from a focus on voluntary prayer in school and tuition tax credits to excellence in education— a key position that Reagan used successfully in his campaign against Walter Mondale.

Like many campaign promises, however, once the votes were tallied, the findings of *A Nation at Risk* were quickly forgotten. Although the report helped to reelect President Reagan and transformed Terrel Bell from the least-known cabinet member at the start of the Reagan presidency into one of the best-known cabinet members by 1984, Bell chose not to serve a second term.

He headed back to obscurity in Utah convinced that America would soon "see the dawn of a new era for American education."[19] So convinced was he that one month before he left office, he claimed in an interview there is "some evidence that we're getting the beginnings of an academic turnaround" in America. His "evidence" was a 0.1 percent increase in the 1984 ACT scores and an even more anemic 0.004 percent increase in the SAT scores.

Politicians conveniently forgot the findings of *A Nation at Risk*. But others didn't.

The Fifth Anniversary of *A Nation at Risk*

On the fifth anniversary of the report's release, Secretary of Education William Bennett said the following:

> American education has made some undeniable progress in the last few years. We are doing better than we were in 1983. But we are certainly not doing well enough, and we are not doing well enough fast enough.

The absolute level at which our improvements are taking place is unacceptably low. We are still at risk. . . . The dropout rate, the poor education of those who do graduate from high school, the widely varying quality of school curriculums, and the manner of promoting teachers and the principals make excellence a matter of chance, not design.[20]

An editorial in the *St. Petersburg Times* noted that "teacher salaries have increased by roughly 40% since 1983" (the year *A Nation at Risk* was released).[21]

The Tenth Anniversary Sees "Wobbly" Results

In 1993, the tenth anniversary of *A Nation at Risk*, an article in the *Washington Post* claimed that "at the time of its release, most figured that this 64-page report would have its day in the headlines and start collecting dust on Washington shelves. Instead, it turned the quality of public education into a major domestic political issue and set in motion forces on which the Clinton administration is building its new education policies."[22]

The *Times-Picayune* in New Orleans wrote an editorial that said "the roller coaster of school reform has bumped along a wobbly track." Barbara Kantrowitz wrote in *Newsweek* that "ten years after a national indictment of American education, the system, consumers, and taxpayers who foot the bill remain dissatisfied." And *USA Today* reported that "few schools have achieved over the past 10 years the dramatic academic improvements evoked in *A Nation at Risk*, despite a flurry of events to beef up classwork."[23]

Critics Raise Their Voices

Everyone was not convinced of the basic premise of *A Nation at Risk*.

Susan Ohanian, a math teacher and an education pundit, wrote, "I wish I had a dime for every media pundit who continues to label *A Nation at Risk* a 'landmark document.' The report was nothing more than a political-industrial-university coterie blowing hot air on the uneasy Zeitgeist. *A Nation at Risk* reflects not a premise for educational change, but a mood of self-justification."[24]

A Popular Indoor Sport

Ohanian's opinion was shared by others. In 1996, David Berliner and Bruce Biddle published a controversial book, *The Manufactured Crisis*, which made four broadside claims about U.S. education progress since "A Nation at Risk" was published:

1. There never was a test score decline.
2. Today's students are outachieving their parents substantially.

3. U.S. students perform well on international assessment tests.
4. The general education malaise crisis in America is a fabrication.

Moreover, in a swipe at report writer Holton, Berliner and Biddle claim that "the bashing of public education has long been a popular indoor sport in America, but never before had criticism appeared that packaged its messages in such flamboyant prose."[25]

Don't Blame the Teachers, It's the Kids' Responsibility

On the tenth anniversary of *A Nation at Risk*, Albert Shanker, the president of the American Federation of Teachers, made a profound observation about the report in his column in the union's magazine:

> *A Nation at Risk* was revolutionary in another way. At the time it appeared, and since, many reports and studies have been fixated on teacher accountability for student achievement, ignoring the fact that in countries with successful school systems, students, not teachers, are held accountable for their achievement. Education reforms are useless unless our kids take responsibility for their education, the way students in other countries do.[26]

Remember these comments; they, too, will appear in the Epilogue as a major recommendation for improving the American public school system.

In 2003, on the twentieth anniversary of the release of *A Nation at Risk*, there was considerable media coverage of the influence of the report—especially since the anniversary occurred shortly after the first anniversary of the signing into law of the No Child Left Behind Act, whose panacea for all educational ills was controversial mandatory testing.

The *Detroit News* wrote, "In the 20 years since *A Nation at Risk*, unions have only supported changes that don't upset the existing system, which is the source of their money and their members' security and power. Why have two decades of reform been so disappointing? A big part [of the answer] rests with the political power of the teacher unions."[27]

The "Power of Persuasion" Is Not Enough

At the twentieth anniversary of *A Nation at Risk*, the commissioners who produced the report were asked to share their observations on its effect. Milton Goldberg, the executive director of the commission, said, "Have we made progress? The answer is yes. Have we accomplished the mandate? The answer is no. I'd say we're a nation that has recognized the risk and is putting in place ways of addressing it."[28]

Gerald Holton, the commission member who wrote the report, lamented, "We didn't have the power of the purse, and the recommendations lacked an implementing authority. We only had the power of persuasion." Norman Francis, now president of Xavier University in Louisiana, lamented, "Our schools are still in trouble. I am not sure the school systems understood the depth of the commitment required for the revolutionary changes we were recommending. If you are not vigilant, committed, passionate, change is not going to happen. And it didn't happen."[29]

A Joke or a Counterculture Antidote?

Some people offered harsh critiques. Gerald W. Bracey, a research psychologist, wrote that "overall *A Nation at Risk* was a grand April Fool's Joke. No Child Left Behind shows we haven't learned a thing in 20 years."[30] Dianne Ravitch, one of the nation's most respected education observers and the assistant secretary of education in the George H. W. Bush administration, summed up what many believed when she wrote, "A product of its time, *A Nation at Risk* proved to be an antidote to many of the pedagogical fads of the 1960s, such as classrooms without walls, fluffy electives, and watered-down curricula that generated public skepticism."[31]

The Twenty-Fifth Anniversary: A Nation of Ostriches?

On the report's twenty-fifth anniversary in 2008, public reaction to *A Nation at Risk* continued unabated. The Department of Education issued a special report in which it emphatically claimed the following:

> If we were "at risk" in 1983, we are at even greater risk now. The rising demands of our global economy, together with demographic shifts, require that we educate more students to higher levels than ever before. Yet our education system is not keeping pace with these growing demands. We simply cannot return to the "ostrich approach" and stick our heads in the sand while grave problems threaten our education system, our civic society, and our economic prosperity. We know which areas need the most attention. Now we must dedicate ourselves to making sure they get it. Twenty-five years after *A Nation at Risk*, can we expect more of our education system? Shouldn't we?[32]

Less than four months later, Paul Houston, then the executive director of the American Association for School Administrators, pushed back hard on the plethora of negative reports about American education launched after *A Nation at Risk* was released in 1983. He told *USA Today* that "the negative reports were the 'rising tide' we got engulfed with—the rising tide of nega-tive reports." He saw the report as "an overstatement of the problem, and it led to sort of hysterical responses. For one, it took liberties with the link between economic development and overall education rates. Yes, the connec-tion makes intuitive sense—but when the dot-com boom made millionaires of ordinary Americans in the 1990s, no one came to my office [Houston was then superintendent of Princeton public schools] and thanked me."[33]

A reader calling herself "Advocate Mom" responded to Houston's newspaper comments by writing that "the year *A Nation at Risk* was released, my husband and I served on a school committee to evaluate and report on how our school district was performing in relation to the concerns of the report. We found many ways in which our own school district had begun the path to mediocrity."[34]

Newsweek said, "After a vast political and financial investment spanning two and a half decades, we [Americans] are far from achieving the report's ambitious aims. But if achieving *A Nation at Risk*'s vision is becoming increas-ingly difficult, the alternative is really no alternative."[35]

You Be the Judge: Read the Report

Just what is this report that has created so much controversy in the past 30 years? Readers can find the entire version of *A Nation at Risk* at http:// datacenter.spps.org/uploads/SOTW_A_Nation_at_Risk_1983.pdf. You de-cide: Was *A Nation at Risk* an overstatement of the problem or a prescient document that was a clarion call to fix the American education system?

Challenge Your Colleagues

Read the opening paragraphs to your colleagues (be sure to omit the refer-ence to *Sputnik*—that's a dead giveaway) and ask them when this was written and by whom.

Most will respond with answers like Tom Friedman's *The World Is Flat* or a recent editorial in a prominent newspaper. All will be surprised when you tell them the document is 30 years old.

1985: *Global Competition: The New Reality*

> Our ability to compete in world markets is eroding. Growth in U.S. productivity lags far behind that of our foreign competitors. Our world leadership is at stake.
>
> —PRESIDENT'S COMMISSION ON INDUSTRIAL
> COMPETITIVENESS, *GLOBAL COMPETITION*[36]

John A. Young, the former chairman of the Hewlett-Packard Corporation, chaired the President's Commission on Industrial Competitiveness,
which produced *Global Competition: The New Reality*, a 1985 report for
President Reagan. This report echoed many of the themes of *A Nation at
Risk*:

Those of you who were around some 25 years ago can remember
what we felt at that very visible image of a Russian rocket blasting its
way into space. That first *Sputnik* wounded our pride, strengthened
our resolve, and set off a national effort to be the first to the moon.
And, of course, we were. What this country needs today is to have the
Japanese launch a Toyota into space. Or perhaps a Sony Walkman.
The competitive challenge we face today has consequences just as
serious as the threat we felt a quarter of a century ago. This one is
just subtler, and a whole lot quieter. Perhaps that is why this nation
has not yet responded wholeheartedly or effectively to the challenge
of competition from abroad.[37]

Although the commission did not specifically refer to the improvement of
the U.S. education system as a solution to the problem of global competitiveness, it did have the foresight to state that the "development of a more skilled
workforce" would be the key to America's future. This theme is repeated often
in future reports.

1985: *Corporate Classrooms: The Learning Business*

Corporate classrooms should not be so busy teaching the 3Rs.
—NELL EURICH, *CORPORATE CLASSROOMS*[38]

One of the primary beneficiaries of America's education system is business, which hires, trains, and (we hope) retains workers who have graduated
from that system. To supplement public education, businesses have for decades deployed corporate education programs to train workers in new skills
necessary to perform specific jobs or to reinforce basic education skills not
learned well in the classroom.

Until 1985, when the Carnegie Foundation for the Advancement of
Teaching released a report entitled *Corporate Classrooms: The Learning
Business*, no one knew the enormous public cost of these corporate learn-
ing programs.

The University of Corporations

In 1985, the corporate education market, at $60 billion, was three times larger than the U.S. Department of Education's $18.9 billion budget.[39] Four years earlier, Harold Hodgkinson, the former director of the National Institute of Education, claimed that the "size and value [of the corporate learning business] was coming close to the net worth of the 3,500 colleges and universities whose total investment is about $55 billion."[40]

Why Are Corporate Programs Necessary?

Why was so much money being spent on education by business? An article published at the report's release echoed the necessary skills premise of the *Corporate Classrooms* report, stating that "the key reason corporations are spending so much to train and educate their workers is that traditional schools, from kindergarten to college, too often produce workers lacking basic communication and problem-solving skills."[41] This "lack of skills" lament will be repeated often.

An Incompetence Tax

The *Corporate Classrooms* report stated emphatically that the American public was underwriting the cost of corporate education. Companies take a tax write-off of 50 percent of the cost of corporate education, and consumers often pay the other 50 percent through higher prices for goods and services sold by the companies.

How have corporate training programs fared in the last three decades? From 1981 to 2011, the amount of money spent by business on employee education rose 287 percent, to $172 billion in 2010, according to the American Society of Training and Development.[42] The U.S. Department of Education's 2011 budget was only $70 billion.

The $172 billion invested in corporate training and education in 2010, in my opinion, represents an annual incompetence tax that Americans must pay because our public schools are not delivering the goods.

1986: *A Nation Prepared: Teachers for the 21st Century*

> As the pool of educated and skilled people in America grows smaller, the backwater of the unemployable rises.
>
> —CARNEGIE FORUM ON EDUCATION AND
> THE ECONOMY, *A NATION PREPARED*[43]

Fourteen years before the turn of the twenty-first century, *A Nation Prepared: Teachers for the 21st Century* was released by the Carnegie Forum on Education and Economy. This report was noteworthy because it was the first to "draw America's attention to the link between economic growth and the skills and abilities of the people who contribute to that growth."[44]

The report also demonstrated the importance of a strong American education infrastructure to the economic vitality of the country with this statement:

> Americans have not yet fully recognized two essential truths: first, that success depends on achieving far more demanding educational standards than we have ever attempted to reach before, and second, that the key to success lies in creating a profession equal to the task—a profession of well-educated teachers prepared to assume new powers and responsibilities. Without a profession of high skills, capabilities, and aspirations, any reforms will be short-lived.[45]

1987: *Workforce 2000: Work and Workers for the Twenty-first Century*

> The educational standards that have been established in the nation's schools must be raised dramatically. Put simply, students must go to school longer, study more, and pass more difficult tests covering more advanced subject matter.
>
> —WILLIAM B. JOHNSON AND ARNOLD
> H. PACKER, *WORKFORCE 2000*[46]

In 1987, the Hudson Institute's *Workforce 2000: Work and Workers for the 21st Century* warned Americans of the transformation the country was confronting as the global economy shifted from a manufacturing-based economy to a service-based economy.

Workforce 2000 left no doubt America's economic future was linked to service jobs:

> U.S. manufacturing will be a much smaller share of the economy in 2000 than it is today. Service industries will create all of the new jobs, and most of the wealth, over the next 13 years. The new jobs in service industries will demand much higher skill levels than the jobs of today. Very few new jobs will be created for those who cannot read, follow directions, and use mathematics.

A century ago, a high school education was thought to be superfluous for factory workers, and a college degree was the mark of an academic or a lawyer. . . . For the first time in history, between now and the year 2000, a majority of all new jobs will require postsecondary education. Education and training are the primary systems by which the human capital of a nation is preserved and increased.[47]

The insightful *Workforce 2000* recommendations, particularly the recommendation about the critical importance of postsecondary education, should have been included by Vannevar Bush in his 1945 report to President Truman, *Science: The Endless Frontier.* But he chose instead to focus national attention on the human capital needs of the nation's elite. If the Hudson Institute's idea of broad-based human development needs had become the goal of American education in 1945, how much further along would we be now?

1987: *The National Science Foundation Annual Report Introduces STEM*

The National Science Foundation is an independent federal agency that was founded in 1950 to "to promote the progress of science; to advance the national health, prosperity, and welfare; and to secure the national defense."[48] By law, it must submit an annual report to Congress and the president summarizing the key trends in science and technology.

According to its 1987 annual report, "our schools and colleges remain far from capable of either teaching the substance of elementary science and mathematics to all students or of preparing students to deal rationally with the social, economic, or philosophical consequences of all this new knowledge."[49]

"Uninspired, Tedious, and Dull"

The main theme of this report was borrowed from a March 1987 National Science Foundation report entitled *Undergraduate Science, Mathematics, and Engineering Education.* The seminal conclusion of that report warned that "while the quantity of United States educational opportunities is high, the quality of that education needs attention. Science, mathematics, and engineering instruction was too often uninspired, tedious, and dull," and education facilities in the United States were "obsolete and inadequate."

The March 1987 report claimed that although "the public is interested in science and technology, it knows little" about those subjects. And even though "the United States leads the world in mathematics research," our country's position in mathematics is "in a precarious situation, and we believe that it may be worsening."[50]

The Godfather of STEM

The most noteworthy legacy of the March 1987 National Science Foundation report was that it was the first report to focus national attention on the importance of science, technology, engineering, and math (STEM) education.

1987: *The Fourth R: Workforce Readiness, a Guide to Business Education Partnerships*

Without fundamental changes and major improvements in the way schools prepare young people, employers will pay an increasingly high price for the educational deficit.

—NATIONAL ALLIANCE OF BUSINESS, *THE FOURTH R*[51]

Shift Happens

The Fourth R: Workforce Readiness, a Guide to Business Education Partnerships, a 1987 report from the National Alliance of Business, added to the national discussion about the shift from a manufacturing-based economy to a service-based economy reported in *Workforce 2000*. *The Fourth R* took direct aim at America's public school system and warned about the critical need to adapt to that shift.

The Need for the Fourth R

The Fourth R states as follows:

Business needs competent, job-ready new workers who have acquired these basic skills (the 3 Rs—reading, writing, and arithmetic), but to be productive today and tomorrow, business needs more. Business seeks young workers with the "Fourth R," workforce readiness, which includes thinking, reasoning, analytical, creative, and problem-solving skills and behaviors such as reliability, responsibility and responsiveness to change and to new work requirements. We do not suggest that schools are the source or locus of all problems or that business has all the solutions. Workforce readiness is a matter of both quantity and quality. The dwindling number of (qualified) workers will require that we look among groups of individuals previously ignored and often considered less ready to work. There are certainly problems with our educational system when a high proportion of young people drop out before completion (10% in 2008), and many of those who do complete their education, lack the basic skills necessary to succeed in the workplace.[52]

The Challenge to Business Leaders: Get Involved

The most provocative word in *The Fourth R*'s title is *guide*. Rather than simply critique the American education system, the National Alliance of Business went further and issued this challenge to business leaders:

> The National Alliance of Business challenges businesses not involved with education to get involved. Businesses already involved must analyze their level of involvement with education and escalate and expand their investments and partnerships, favoring those which bring about systemic educational improvements and policy changes. Partnerships are essential at all levels—kindergarten through grade 12—including alternative education programs, vocational education, remediation and literacy programs, and preschool programs. These partnerships are a means for businesses to address their own workforce needs and their community's social policy agenda at the same time.[53]

The Big Picture Develops

One of the benefits of writing a book like this is how much you learn. Before we move on, I think it is important to point out the pattern that is developing: a widespread recognition in the mid-1980s, mostly by business and government, that something new was happening in and to America—an inextricable global movement away from a manufacturing-based economy to a service-based one.

A Nation at Risk was perhaps the first such national warning that a shift in skills was occurring. But to understand more fully the big picture about the state of American education in the twenty-first century, one has to take into account that for the past 30 years, a lot of smart people, smart associations, and smart politicians have warned that this shift not only was coming but was already here.

Even in the early 1950s, there was a fierce debate among academics about how and what to teach. I am not advocating a total restructuring of the American curriculum system in 2013, but I am recommending strongly that someone take the lead in the American education system and, starting at the third-grade level, integrate skills like collaboration, analytics (yes, even at this age level), and critical thinking into our national curriculum.

The future of America's economy, workforce employment, and national security depends on how well we do this.

1989: *Winning the Brain Race: A Bold Plan to Make Our Schools Competitive*

> Human history becomes more and more a race between education and catastrophe.
>
> —**H. G. Wells**[54]

This quote was chosen by David T. Kearns, the former chairman and CEO of the Xerox Corporation, and Denis P. Doyle, an education research fellow at the Hudson Institute, to open their 1989 book, *Winning the Brain Race: A Bold Plan to Make Our Schools Competitive*. The book continued the dialogue among national leaders on the imperative of preparing the future American workforce for a competitive environment that values brains more than brawn.

Not My Problem

As CEO of Xerox, Kearns had a front-row seat in the 1970s and 1980s as Xerox competed for market share with Asian (mostly Japanese) corporations in the office automation, printing, and copier markets. From that brutal global competition, Kearns learned that the most important strategy for improving the nation's competitive response to global competition was a better education system.

Frustrated that others knew this, too, but that little was being done, he wrote, with his coauthor Doyle, "We are convinced that the nation faces an education crisis. And most Americans agree. Yet too little is being done. Why? Because Americans believe the education crisis is someone else's problem."[55]

Coasting Rather Than Competing

Kearns and Doyle offered this analysis of the American business climate in 1989:

> American business let its guard down in the 1970s. Coasting on a reputation earned during and after World War II, American manufacturers thought they could do no wrong. In business and commerce, the race

goes to the swift, and the penalty for losing has been harsh and painful. We have lost not only market share but whole industries to foreign competitors. America invented and used to own electronics, television, high-performance automobiles, facsimile transmission, the computer chip, PCs . . . the list is long and dreary.[56]

The list began in August 1950, when 75 Japanese CEOs listened to, and then embraced, the total quality management advice from W. Edwards Deming. American leaders could have adopted that advice at the time, too, but they didn't.

Restructure the Whole System

What should be done? Kearns and Doyle believed that nothing less than a wholesale restructuring of the American education system was in order:

> Public education in this country is in crisis. America's public schools graduate 700,000 functionally illiterate students every year, and 700,000 more drop out. Four out of five young adults couldn't summarize the main point of a newspaper article, read a bus schedule, or figure their change from a restaurant bill. The task before us is the restructuring of our entire public education system. We don't mean tinkering. We don't mean piecemeal changes or well-intentioned reforms. We mean a total restructuring of our schools.[57]

Kearns and Doyle offer a six-point plan for the "total restructuring" of the American education system:

1. **Choice.** Public schools should compete with one another; students and teachers should be able to choose the schools they want to go to.
2. **Restructuring.** Schools should be open all year round and be run by teachers and principals.
3. **Professionalism.** Teachers must raise standards to elevate their status to the level of other professionals.
4. **Standards.** Academic standards must be raised and children held strictly accountable to them; it is the student's job to learn—no promotion without performance.
5. **Values.** We are producing a generation of young Americans that neither appreciates nor understands our democratic society.
6. **Federal responsibility.** The federal government's role in education is limited and should continue to be so, but the federal government should do more than it does, particularly in research.[58]

Business Has to Set the Agenda

Although the restructuring plan outlined in *Winning the Brain Race* renewed the urgency to get America's business leaders involved in improving the country's education system, some in the business community thought the ideas were too broad and too radical. Kearns and Doyle took on their critics with these words:

> Despite its interest in education, the business community has been disappointingly unspecific about education reform. Critics of public education can no longer enjoy the luxury of criticism without accepting responsibility for suggesting ways to change the system. Business and education have largely failed in their efforts to improve the schools, because education set the agenda. Driven by competition and market discipline, business will have to set the new agenda.[59]

A Joint Agenda Is Needed

To me, the most profound comment in the book is that *business* must set the "new agenda" for public education in America.

Nearly a quarter century after Kearns and Doyle made that proclamation, the education community—teachers, school administrators, and teacher unions—continue to set our nation's education agenda. Although more and more businesses are concerned and involved in educational issues, few have really made the improvement of the U.S. education system a corporate priority. The reason for this is that "long-term" for business leaders in America is the next 90 days. A "total restructuring" of America's education system cannot happen in 90 days, 900 days, or even 9,000 days. Moreover, it cannot happen without both groups—business leaders and education leaders—being engaged, committed, and responsible for the agenda and for the outcomes.

Notes

1. Myron Lieberman, *The Future of Public Education* (University of Chicago Press, Chicago, 1960), 6.
2. International Association for the Evaluation of Educational Achievement, First International Mathematics Study, 1964, www.iea.nl/fims.html.
3. International Association for the Evaluation of Educational Achievement, First International Science Study, 1971, www.iea.nl/fiss.html.
4. Edward Simon, "Governor Milton J. Shapp's Proposal for a National Education Trust Fund," Boston Federal Reserve Bank, June 1971, www.bos.frb.org/economic/conf/conf7/conf7k.pdf.
5. National Center for Education Statistics, *The Nation's Report Card*, 1978, www.nces.ed.gov/nationsreportcard/pdf/main2008/2009479.pdf.

6. Ibid.
7. F. Fowler and T. Poetter, *Framing French Success in Elementary Mathematics Curriculum: Implications for American Educators* (Montreal: American Educational Research Association, 1999).
8. International Association for the Evaluation of Educational Achievement, *Second International Mathematics Study*, 1983, www.iea.nl/sims.html.
9. K. M. Hart, *Educational Studies in Mathematics*, 19, no. 2 (May 1988): 273.
10. Gerald Holton, *A Nation at Risk: The Imperative for Educational Reform* (Washington, DC: U.S. Department of Education, 1983), 9, http://datacenter.spps.org/uploads/SOTW_A_Nation_at_Risk_1983.pdf.
11. Terrel H. Bell, *The Thirteenth Man: A Reagan Cabinet Memoir* (New York: Free Press, 1988), 2.
12. Bret Schulte, "Ronald Reagan v. Jimmy Carter: Are You Better Off Now Than You Were Four Years Ago?," *U.S. News & World Report*, January 17, 2008.
13. Bell, *The Thirteenth Man*, 114.
14. Ibid., 115.
15. Ibid., 117.
16. Christopher Connell, "Education Panel Sees 'Rising Tide of Mediocrity,'" Associated Press, April 27, 1983; "U.S. Education: Unsatisfactory," *Miami Herald*, May 8, 1983; Admiral H. G. Rickover, "Failure in Education," *Washington Post*, June 19, 1983; and Dick Pothier, "Teaching Teachers: Many Think It's Key to Better Schools," *Lexington Herald-Ledger*, September 4, 1983.
17. Gerald W. Bracey, "April Foolishness: The 20th Anniversary of a Nation at Risk," *Phi Delta Kappan* 84, no. 8 (2003): 616.
18. Harold Jackson, "Reagan Calls Moscow an Evil Empire," *Guardian*, March 9, 1983.
19. Bell, *The Thirteenth Man*, 127.
20. "Bennett Reports U.S. Education Still at Risk," Associated Press, April 25, 1988.
21. "A Nation (Still) at Risk," *St. Petersburg Times*, September 12, 1988.
22. Edward W. Fiske, "The Report That Accidentally Shook Up Schools," *Washington Post*, May 4, 1993.
23. Pamela Coyle, "Ten Years Later: A Nation Still at Risk?," *Times-Picayune*, April 1, 1993; Barbara Kantrowitz, "A Nation Still at Risk," *Newsweek*, April 19, 1993; and Tamara Henry, "Results below Expectations, Analyzing the Lack of True Gains," *USA Today*, March 17, 1993.
24. Susan Ohanian, "A Sham, an Insult, the 'Nation at Risk' Study Was Nothing More Than Finger Pointing by Those Who Can't Teach," *San Jose Mercury News*, May 16, 1993.
25. David C. Berliner and Bruce J. Biddle, *The Manufactured Crisis: Myths, Fraud, and the Attack on America's Public Schools* (Boston: Addison Wesley Longman, 1995), 139.
26. Albert Shanker, "Where We Stand," *Education Next*, May 9, 1993.
27. Terry M. Moe, "Teacher Unions Delay Education Reforms," *Detroit News*, June 10, 2003.
28. Ben Feller, "School Results Mixed 20 Years after 'Risk' Report," Associated Press, April 26, 2003.
29. Joan Ryan, "We Are a Nation Still at Risk," *Seattle Post-Intelligencer*, April 15, 2003.
30. Gerald W. Bracey, "We Were Never at Risk," *Cincinnati Post*, April 29, 2003.
31. Diane Ravitch, "A Nation at Risk," *Trenton Times*, March 24, 2003.
32. U.S. Department of Education, *A Nation Accountable: 25 Years after "A Nation at Risk"* (Washington, DC: U.S. Government Printing Office, 2008), 1.
33. Gregg Toppo, "Nation at Risk: The Best Thing or the Worst Thing for Education?" *USA Today*, August 1, 2008.
34. Ibid., http://usatoday30.usatoday.com/news/education/2008-04-22-nation-at-risk_N.htm.
35. Thomas Toch, "Still at Risk: 25 Years of U.S. Education; Reform Still Needed," *Newsweek*, April 24, 2008.
36. President's Commission on Industrial Competitiveness, *Global Competition: The New Reality* (Washington, DC: U.S. Government Printing Office, 1985), 1.

37. Ibid.
38. Nell P. Eurich, *Corporate Classrooms: The Learning Business* (Lawrenceville, NJ: Princeton University Press, 1985), 15.
39. "U.S. Department of Education," August 5, 2011, www.ed.gov/about/overview/budget/history/edhistory.pdf.
40. Eurich, *Corporate Classrooms*, 20.
41. "School, College Failings Cost Business Billions," Associated Press, January 29, 1985.
42. American Society of Training and Development, *2011 State of the Industry Report* (Alexandria, VA: American Society of Training and Development, November 1, 2011), www.astd .org/Publications/Magazines/TD/TD-Archive/2011/11/The-2011-State-of-the-Industry-Increased-Commitment-to-Workplace-Learning.
43. Carnegie Forum on Education and the Economy, *A Nation Prepared: Teachers for the 21st Century* (New York: Carnegie Forum on Education and the Economy, 1986), 2.
44. Ibid.
45. Ibid.
46. William B. Johnson and Arnold H. Packer, Workforce 2000: Work and Workers for the 21st Century (Washington, DC: Hudson Institute, 1987), 117.
47. Ibid., 107.
48. National Science Foundation, *Annual Report 1987* (Washington, DC: U.S. Government Printing Office, 1987), 7.
49. Ibid., 7.
50. Ibid., 8.
51. National Alliance of Business, *The Fourth R: Workforce Readiness: A Guide to Business Education Partnerships* (Washington, DC: U.S. Department of Labor, 1987), 6.
52. Ibid., 7.
53. Ibid.
54. Quoted in Denis P. Doyle and David T. Kearns, *Winning the Brain Race: A Bold Plan to Make Our Schools More Competitive* (San Francisco, CA: Center for Self Governance, 1989), 5.
55. Ibid., 2.
56. Ibid., 3.
57. Ibid., 5.
58. Ibid., 164.
59. Ibid., 163.

CHAPTER 10
The Skills Gap Emerges

To ensure a more prosperous future, we must improve productivity and our competitive position. We cannot simply do this by using better machinery, because low-wage countries can now use the same machines and can still sell their products more cheaply than we can. America is headed toward an economic cliff. If basic changes are not made, real wages will continue to fall, and the gap between economic "haves" and "have-nots" will widen still further.

—COMMISSION ON THE SKILLS OF THE AMERICAN
WORKFORCE, *AMERICA'S CHOICE*[1]

1990: *America's Choice: High Skills or Low Wages!*

The opening quote sounds like it's from a 2013 political speech, doesn't it? But it came from a 1990 report from the Commission on the Skills of the American Workforce, appointed by the National Center on Education and the Economy, a Rochester, New York–based nonprofit group "created to develop proposals for building the world-class education and training system that the United States must have if it is to have a world-class economy."[2]

A New Model Is Necessary

Based on 2,000 interviews with respondents at more than 550 companies and government agencies in seven countries, *America's Choice: High Skills or Low Wages!* reiterated the growing crescendo of warnings: the American workforce and education system were missing the seismic economic shift from a manufacturing-based economy to a service-based economy. The report warned the following:

> The organization of America's workplace today is largely modeled after the system of mass manufacture pioneered during the early 1900s. The premise is simple: break complex jobs into a myriad of simple rote tasks which the worker then repeats with machine-like efficiency. Most employees under this model need not be educated. It is far more important that they be reliable, steady, and willing to follow directions.[3]

America's Choice delivered a blunt appraisal of America's global competitive position:

> The Commission has grown increasingly uneasy as we have watched Singapore, Taiwan, and [South] Korea grow from run-down Third World outposts to world premier exporters; as Germany, with one quarter of our population, almost equaled us in exports; as Japan became the world's economic juggernaut; and, as America became the world's largest borrower.[4]

America's Choice also criticized America's education system:

> Our education statistics are as disappointing as our trade statistics. Our children rank at the bottom on most international tests. Again we hear the excuses. They have elite systems, but we educate everyone. They compare a small number of their best to our much larger average. The facts are otherwise; many of the countries with the highest test scores have more of their students in school than we do.[5]

The Choice Is Ours

The commission offered this solution:

> The key to maintaining or improving our standard of living is productivity growth—more products and services from every member of the workforce. But, during the past two decades, our productivity growth has slowed to a crawl. If productivity continues to falter, we can expect one of two outcomes. Either the top 30 percent of our population will grow wealthier while the bottom 70 percent becomes progressively poorer, or we will all slide into relative poverty together. This is our choice: high skills or low wages.[6]

You can find the entire report here: www.skillscommission.org/?page_ id=296.

1990: The Second International Science Study

In 1990, 19 years after the First International Science Study, the International Association for the Evaluation of Educational Achievement released the results of the Second International Science Study.

As shown in Table 10.1, American eighth graders performed poorly, placing a dismal 13th out of 17 countries in an overall science skills assessment test.[7]

Table 10.1   Mired in Thirteenth Place

Country	Number of Items (Out of 30) Answered Correctly (Eighth Graders)
Hungary	22
Japan	20
Netherlands	20
Canada	19
Finland	19
Sweden	18
South Korea	18
Poland	18
Norway	18
Australia	18
England	17
Italy	17
United States	**16**
Singapore	16
Thailand	16
Hong Kong	16
Philippines	12

1990: The National Assessment of Educational Progress

The Department of Education, through the National Assessment of Educational Progress (NAEP), has administered math and science assessment tests since 1969. In 1990 the department introduced four new criteria-based achievement measurements—categories that remain a source of controversy in 2013.

The four categories introduced were: *advanced*, which represents superior performance (think A); *proficient*, the label given to students who have demonstrated "competency over challenging subject matter" (think B); *basic*, which describes students who exhibit "partial mastery of prerequisite knowledge and skills" (think C); and *below basic*, which is used for students who have not yet attained even partial mastery of subject matter (think D or lower).

Basically, a Poor Showing

Table 10.2 shows the NAEP scores for fourth, eighth, and twelfth-grade American students in mathematics in 1990.[8]

These scores are clearly troubling. One would think that in the two higher grades, where students have a teacher specifically trained in math teaching them that subject, there would be a trend toward proficient and advanced results. That doesn't happen. In fact, the scores for twelfth grade are the worst grade!

Table 10.2   Hard to Get beyond the Basics

	Advanced	Proficient	Basic	Below Basic
Fourth grade	1%	12%	37%	50%
Eighth grade	2%	13%	35%	48%
Twelfth grade	1%	11%	46%	42%

We're Number One—Not Yet!

In the beginning of 1990, President George H. W. Bush proclaimed that American students "will be first in the world in mathematics and science achievement by the year 2000."[9] The results of the 1990 National Assessment of Educational Progress in mathematics indicated that this goal would be difficult to achieve.

1993: John Sculley: "America Is Resource Poor"

Seymour Papert, the founder of the MIT Media Lab and the winner of the Computerworld and Smithsonian Institution's Search for New Heroes award mentioned earlier, published a book in 1993 for which he asked Apple Computer chairman and CEO John Sculley to write the foreword. Considering when this foreword was written, Scully was prescient in his vision of the future of America's economic position and its educational system:

> During the Industrial Age and for most of this century, America stood alone at the top of an economic pyramid, taking resources out of the ground—oil, wheat, coal—adding its manufacturing know-how to those resources, and selling those goods to the rest of the world. We are no longer in the Industrial Age. We are in the Information Economy, where strategic advantage is determined by ideas and information and by the skills of a nation's workforce.
>
> Virtually overnight, America has gone from being resource rich to being resource poor. As a direct consequence, America is perched on the edge of an economic cliff, and unless we make a concerted effort to bring the educational system into sync with the rest of the global economy, we are in danger of supplying the rest of the world with low-wage work and losing out on the high-skill, high-wage economy that the world has moved toward.
>
> Once again you hear the clarion call for a new set of skills to be taught to Americans—skills needed to insure the country's economic, global workforce and national security futures in the post-industrial age.[10]

1995: The Third International Mathematics and Science Study

The Third International Math and Science Study was the largest and most ambitious international study to date of global student achievement in math and science.

Once again, American eighth graders performed poorly, being number 17 out of 18 countries in mathematics (see Table 10.3) and number 13 out of 24 countries in earth science.[11] The 1995 Third International Mathematics and Science Study was the fifth global assessment test. It was also the fifth consecutive such test in which American students lagged far behind the rest of the world. American students were not making progress in responding to President Bush's 1990 challenge to be the best in the world by the year 2000, only five years in the future.

Table 10.3 Still Way Behind

Country	Number of Math Questions (Out of 151) Answered Correctly (Eighth Graders)
Singapore	79
Japan	73
South Korea	72
Hong Kong	70
Belgium (Flemish)	66
Czech Republic	66
Slovak Republic	62
Switzerland	62
Hungary	62
France	61
Russian Federation	60
Canada	59
Ireland	59
Sweden	56
New Zealand	54
Norway	54
United States	**53**
England	53

Different Measurement, Improved Ranking

Four years later, in 1999, the International Association for the Evaluation of Educational Achievement changed the scoring metrics from the number of questions answered correctly to an aggregate score. As Tables 10.4 and 10.5 reveal, American eighth graders made good progress in the rankings in this study, which is administered every four years.

Table 10.4 How U.S. Eighth Graders Performed Compared to the Rest of the World in Mathematics

Year	Rank in Study
1999	18th
2003	15th
2007	10th
2011	9th

Table 10.5 How U.S. Eighth Graders Performed Compared to the Rest of the World in Science

Year	Rank in Study
1999	18th
2003	9th
2007	11th
2011	10th

Full copies of the 1995, 1999, 2003, 2007, and 2011 Trends in Mathematics and Science Study can be found at www.iea.nl/completed_studies.html.

1996: The National Assessment of Educational Progress

The 1996 NAEP study focused on mathematics and grouped student results into four achievement categories: advanced, proficient, basic, and below basic.

Not Exactly Proficient

When you consider that powerful forces in Washington, D.C., were determined to set *proficient* as the minimum national goal, the 1996 results continued to be less than stellar: 79 percent of fourth graders, 76 percent of eighth graders, and 84 percent of twelfth graders were less than proficient in math. Moreover, the weak results continue to reinforce the woeful story that American students perform less well the further they progress in the American education system, when exactly the opposite should be happening (see Table 10.6).

Thorny Question: What to Test?

As mentioned earlier, during this time the Clinton administration recommended compulsory testing, and as the data shows, American students and

Table 10.6 1996 National Assessment of Educational Progress Results

	Advanced	Proficient	Basic	Below Basic
Fourth grade	2%	19%	43%	36%
Eighth grade	4%	20%	38%	38%
Twelfth grade	2%	14%	53%	31%

Source: National Center for Education Statistics, *National Assessment of Educational Progress*, 1996, http://nces.ed.gov/nationsreportcard/pubs/main1996/1999452.asp.

teachers needed it. But *recommending* compulsory testing is far different than implementing such an ambitious and controversial program, which school administrations, teachers, teacher unions, business leaders, parents, and politicians all had different views on.

Secretary of Education Richard Riley, the Clinton administration's point man for building a consensus on the compulsory testing program, ran into roadblocks at every turn. Although many agreed with President Clinton that compulsory testing was necessary (as the scores showed), no consensus was reached about what to test or how to test it. The infamous No Child Left Behind Act addressed this issue in 2001.

1999: *New World Coming: American Security in the 21st Century*

> For many years to come Americans will become increasingly less secure, and much less secure than they now believe themselves to be.
>
> —COMMISSION ON NATIONAL SECURITY IN THE
> 21ST CENTURY, *NEW WORLD COMING*[12]

So opens the 1999 report *New World Coming: American Security in the 21st Century*—better known as the Hart-Rudman Report for the two senators, Gary Hart and Warren B. Rudman, who cochaired the Commission on National Security in the 21st Century. The report offered chilling and eerily accurate security predictions for America's future.

A Nation to Be Less Secure

Nowhere does the report actually address the importance of science and math to our nation's security, but the report's conclusion does underscore the strategic importance of producing a nation of skilled individuals in STEM to America's future.

The commission's remarkable forecast says the following:

1. America will become increasingly vulnerable to attack on our homeland, and Americans will likely die on American soil.
2. Rapid advances in information and biotechnologies will create new vulnerabilities for U.S. security.
3. New technologies will divide the world as well as draw it together.
4. The national security of all advanced states will be increasingly affected by the vulnerabilities of the evolving global economic infrastructure.
5. Energy will continue to have major strategic significance.

6. New technologies will continue to stretch and strain all existing borders.
7. Global connectivity will allow "big ideas" to spread quickly around the globe; some ideas may be religious in nature, some populist, and some devoted to democracy and human rights.
8. States will differ in their ability to seize technological and economic opportunities and establish the social and political infrastructure necessary for economic growth.
9. Space will become a critical and competitive military environment, with weapons likely put in space.
10. Electronic communications will continue to expand intelligence collection capabilities around the world, and the United States' technological superiority will fail to detect all dangers in an ever-changing world.[13]

Read That List Again!

Think 9/11. Twitter. Arab Spring. Think of the inability of the coalition forces to find Osama bin Laden for so long. Think of the 2007 global financial meltdown. The commission's list is quite remarkable for its foresight. For the first time, America's national security has been added to the list of reasons the American education system needs to do a much better job teaching STEM to U.S. students.

Notes

1. Commission on the Skills of the American Workforce, *America's Choice: High Skills or Low Wages!* (Rochester, NY: National Center on Education and the Economy, 1990), 2.
2. Ibid., 3.
3. Ibid., 15.
4. Ibid., 16.
5. Ibid., 17.
6. Ibid., 18.
7. International Association for the Evaluation of Educational Achievement, *Second International Science Study*, 1990, www.iea.nl/siss.html.
8. National Center for Education Statistics, *National Assessment of Educational Progress*, 1990, http://nces.ed.gov/pubsearch/pubsinfo.asp?pubid=2012458.
9. George H. W. Bush, "Goals 2000: Educate America," State of the Union speech, January 31, 1990.
10. John Sculley, "Foreword," in Seymour Papert, *Mindstorms: Children, Computers, and Powerful Ideas* (New York: Perseus Books, 1993), vii.
11. International Association for the Evaluation of Educational Achievement, *Third International Mathematics and Science Study*, 1995, www.iea.nl/timss_1995.html.
12. Commission on National Security in the 21st Century, *New World Coming: American Security in the 21st Century: Major Themes and Implications* (Washington, DC: U.S. Government Printing Office, 1999), 2.
13. Ibid., 4.

CHAPTER 11
The Skills Gap Widens

"When things go wrong, you'll find they usually go on getting worse for some time."

—C. S. LEWIS, *THE CHRONICLES OF NARNIA*[1]

The warnings from 1964 to 1999 about the American education system's failures in math and science were the proverbial calm before the storm. As a new century was ushered in, the frequency and seriousness of the reports increased significantly, reaching new levels as the performance of American students in domestic and international science and math tests continued their significant decline. The result was that the U.S. science and math skills gap was about to widen precipitously.

2000: *Ensuring a Strong U.S. Scientific, Technical, and Engineering Workforce in the 21st Century*

If current trends persist, our nation may not have all the talent it will need to enable the innovation process that has given America a strong economy and high quality of life. There is already evidence that worker shortages are limiting economic growth, and industry has repeatedly called for increases in visa quotas that allow technically skilled non-immigrants to work in the United States.

—NATIONAL SCIENCE AND TECHNOLOGY COUNCIL, *ENSURING A STRONG U.S. SCIENTIFIC, TECHNICAL, AND ENGINEERING WORKFORCE IN THE 21ST CENTURY*[2]

The National Science and Technology Council, a cabinet-level commission, was tasked with reporting recommendations to President Clinton on creating a competent U.S. workforce well-grounded in technology skills.

Coming Up Short

The council's report, *Ensuring a Strong U.S. Scientific, Technical, and Engineering Workforce in the 21st Century*, raised a new issue of national concern: the potential of a workforce shortage of U.S. scientists and engineers, a by-product of the American education system's failure to produce a more diverse science and engineering workforce.

The council said the potential future shortage was caused by four reasons:

1. A dominance by white males, who held 65 percent of all science and engineering jobs.
2. A sporadic embrace of science and engineering jobs by women (higher in biological science, lower in engineering).
3. The percentage of minorities dropping off at successive levels of education from undergraduate to graduate to the workforce.
4. The high dependence of the U.S. technology ecosystem on naturalized U.S. citizens or non-U.S. immigrants.[3]

The H-1B Visa Debate

The report's reference to "the high dependence of the U.S. technology ecosystem on . . . non-U.S. immigrants" addressed a STEM labor topic that continues to be controversial: the H-1B visa, a temporary work program created in 1990 by the United States Citizenship and Immigration Services "to temporarily employ foreign workers in specialty occupations like engineering, physical sciences, and mathematics who possess at least a bachelor's degree or equivalent."[4] In April 2000, the same month that the council's report on ensuring a strong workforce was released, 164,814 foreign nationals applied for the 115,000 available H-1B visas.

Supporters of the program, mostly U.S. tech companies in need of hiring skilled tech workers, claim that the U.S. education system is not producing enough workers with STEM undergraduate or graduate degrees. These companies say they need the foreign workers hired by the H-1B visa program to do business.

Opponents say the H-1B visa is a ruse. Their main claim is that companies hire H-1B workers for one reason: They cost less than American science and technology workers.

Ordinary People?

The most articulate spokesperson for the opponents of the H-1B visa is Norman Matloff, a professor of computer science at the University of California, Davis. The H-1B visa program, according to Matloff, "is fundamentally about cheap, indentured labor. Though the tech industry lobbyists portray the H-1B visa as a remedy for labor shortages, and as a means of hiring the 'best and brightest from around the world'. . . the H-1B is about [hiring] cheap immobile labor."[5]

What America Needs Is Answers

What is the status of the STEM skills shortage in 2013? Does one exist? Those who think not point to multidegreed STEM graduates waiting tables in restaurants, unable to find STEM jobs. On the other side are CEOs who lament that technology job vacancies remain open for months because there are few qualified candidates to hire.

U.S. News & World Report has taken a leadership position in the STEM market by hosting a STEM conference each year in June. Before the 2012 conference, Brian Kelly, the magazine's chief content officer, summed up the current situation with this comment: "We all know there is a STEM skills shortage, so we are not going to rehash old news. It's time to figure out some answers."[6] All sides can at least agree with Kelly that we need answers. The shortage is real. And it has the potential to be a major disruptor to America's future economy.

2000: *Before It's Too Late*

> It is abundantly clear from the evidence at hand that we are not doing the job that we should do, or can do, in teaching our children to understand and use ideas from science and mathematics. Our children are falling behind; they are simply not world-class learners when it comes to science and mathematics.
>
> —NATIONAL COMMISSION ON MATHEMATICS AND SCIENCE
> TEACHING FOR THE 21ST CENTURY, *BEFORE IT'S TOO LATE*[7]

Also known as the Glenn Commission after its chairman, Senator John Glenn, the National Commission on Mathematics and Science Teaching for the 21st Century was a 25-person task force empowered by Secretary of Education Richard Riley in 2000 to "investigate and report on the quality of mathematics and science teaching in the nation and to consider ways of improving recruitment, preparation, retention and professional growth for K–12 mathematics and science teachers."[8]

Out of Their Field

The report, *Before It's Too Late*, said that "the teaching pool in mathematics and science is inadequate to meet our current needs; many classes in these subjects are taught by unqualified and underqualified teachers."[9] Four years after the report's release, PhD candidate Anne Kolarik focused on the qualifications of mathematics and science teachers addressed in the report. A key criterion she used in her analysis was the term called *out of field*. A mathematics or science teacher is out of field when he or she does not have an undergraduate degree in the subject. Kolarik found that from 2000 to 2004, out-of-field teaching in the United States worsened in mathematics but improved in science (see Table 11.1).[10]

Table 11.1 Math and Science Teachers with No STEM Degree

	2000 Teaching Out of Field	2004 Teaching Out of Field
Math	71%	79%
Science	72%	60%

Let's Do It!

The most innovative recommendation of *Before It's Too Late* was to create 15 mathematics and science teaching academies to train 3,000 teaching fellows in effective teaching methods in mathematics and science in a one-year intensive course. The report explains as follows:

> The three thousand fellows will be identified from recent college graduates and people at midcareer who are seeking a new challenge in life. Each fellow will be appointed to receive a one-year, intensive course in effective teaching methods in mathematics and science and receive a $30,000 stipend. After the program the fellows will be ideal job candidates for school districts suffering from mathematics and science teacher shortages in middle school and high school.[11]

Thirteen years later, this proposal remains an extraordinary idea. It has the potential to significantly upgrade the quality of math and science education in America, and it is one of the ideas I recommend in the Epilogue.

2000: The Programme for International Student Assessment

A new international math and science assessment test was introduced in 2000. This test, sponsored by the Organization for Economic Cooperation and Development (OECD) and slated to be administered every three years, is called the Programme for International Student Assessment (PISA). It has a specific measurement target: the math and science skills of 180,000 tenth-grade students around the world.

Still in the Tracks of Their Peers

Similar to the results described in the Trends in Mathematics and Science Study, America's tenth graders performed poorly in the initial PISA test in both math and science, being number 19 in mathematics and number 14 in science out of 30 countries (see Tables 11.2 and 11.3).[12]

Table 11.2 Number 19 in Math

Country	Math Literacy Score
Japan	557
South Korea	547
New Zealand	537
Finland	536
Australia	533
Canada	533
Switzerland	529
United Kingdom	529
Belgium	520
France	517
Austria	515
Denmark	514
Iceland	514
Liechtenstein	514
Sweden	510
Ireland	503
OECD Average	500
Norway	499

(Continued)

Table 11.2   (*Continued*)

Country	Math Literacy Score
Czech Republic	498
United States	**493**
Germany	490
Hungary	488
Russia	478
Spain	476
Poland	470
Italy	457
Portugal	454
Greece	447
Luxembourg	446
Mexico	387
Brazil	334

Table 11.3   Number 14 in Science

Country	Science Literacy Score
South Korea	552
Japan	550
Finland	538
United Kingdom	532
Canada	529
New Zealand	528
Australia	528
Austria	519
Ireland	513
Sweden	512
Czech Republic	511
France	500
Norway	500
OCED average	500
United States	**499**
Hungary	496
Iceland	496
Belgium	496

Table 11.3   (*Continued*)

Switzerland	496
Spain	491
Germany	487
Poland	483
Denmark	481
Italy	478
Greece	461
Russia	460
Portugal	459
Luxembourg	443
Mexico	422

The Newer the Test, The Worse the Result

The PISA science and mathematics examination, conducted every three years, was the newest test to deliver a familiar result: American students, this time 10th graders, continue to lag far behind their global peers in math and science skills. Listed below, in Tables 11.4 and 11.5, are the rankings of American students (out of 41 countries in 2003, 57 in 2006, and 74 in 2009) in the 2003, 2006, and 2009 PISA examinations. As you can see, American tenth-grade students have performed worse with each year.

Full copies of the results of the PISA examinations can be found at www .oecd.org/pisa/pisaproducts.

Table 11.4   Ranking for U.S. Tenth Graders in Mathematics

Year	U.S. Ranking/No. of Participating Countries
2003	23/41
2006	25/57
2009	31/74

Table 11.5   Ranking for U.S. Tenth Graders in Science

Year	Ranking
2003	14
2006	21
2009	23

2000: The National Assessment of Educational Progress Test

After an eight-year political struggle by the Clinton administration to make testing compulsory in the United States, the National Assessment of Educational Progress (NAEP) Test was administered to America's fourth, eighth, and twelfth graders in 2000. American students continued their struggle to be rated proficient in math and science. This was the last NAEP test to be conducted before compulsory testing became the law of the land in the George W. Bush administration. See Tables 11.6 and 11.7 for the results.[13]

Once again, the math and science skills of American students only became worse as the students progressed through the education system.

Curiosity

As I reviewed the 2000 NAEP data, I was curious about the following: How do the percentages for twelfth graders compare to the results of the 1941 Army General Classification Test (discussed earlier)? My gut feeling was that the data would be uncannily similar.

The military used the Army General Classification Test to classify recruits in five categories: (1) fast learners, (2) an unnamed category, (3) average learners, (4) below average learners, and (5) slow learners. Those in categories 4 and 5 were considered to have the mental capacity of third graders.

To make the comparison between the 1941 data and the 2000 NAEP data, I assigned the advanced NAEP category to the army's fast learners, the proficient NAEP category to the army's unnamed second category, the basic

Table 11.6   Science Scores: Percent Rated

Grade	Advanced	Proficient	Basic	Below Basic
Fourth	4%	26%	37%	34%
Eighth	4%	28%	29%	39%
Twelfth	2%	16%	34%	47%

Table 11.7   Mathematics Scores: Percent Rated

Grade	Advanced	Proficient	Basic	Below Basic
Fourth	3%	23%	43%	31%
Eighth	5%	22%	38%	34%
Twelfth	2%	15%	48%	35%

Table 11.8 2000 NAEP versus 1941 Army Science Scores

Testing Group	Advanced	Proficient	Basic	Below Basic
NAEP 2000	2%	16%	34%	47%
Army 1941	6%	26%	30%	38%

Source: NAEP 2000 scores and Army General Classification Test 1941 scores

Table 11.9 2000 NAEP versus 1941 Army Math Scores

Testing Group	Advanced	Proficient	Basic	Below Basic
NAEP 2000	2%	14%	48%	35%
Army 1941	6%	26%	30%	38%

Source: NAEP 2000 scores and Army General Classification Test 1941 scores

NAEP category to the army's average learners, and the below basic NAEP category to the army's below average and slow learners.

Tables 11.8 and 11.9 show what I found.[14]

Our Ancestors Were Smarter

I was wrong. In my unscientific comparison, the 1941 inductees to the U. S. military performed better than their descendants did 59 years later! Granted, my approach was subjective.

But it does support a basic premise of this book: For decades, Americans have not mastered the subjects of science and math or performed well in general aptitude tests.

And just as I was wrong, so, too, were David Berliner and Bruce Biddle, who claimed in their 1996 book, *The Manufactured Crisis*, that "today's students were outachieving their parents substantially."[15]

2002: *Unraveling the Teacher Shortage Problem: Teacher Retention Is the Key*

"Our inability to support high-quality teaching is driven not by too few teachers coming in but by too many going out; that is, by a staggering teacher turnover and attrition rate.

—NATIONAL COMMISSION ON TEACHING AND AMERICA'S FUTURE,
UNRAVELING THE TEACHER SHORTAGE PROBLEM[16]

The National Commission on Teaching and America's Future (NCTAF) is a nonprofit organization whose mission is to improve the overall quality of classroom teachers. Much of the debate about teacher quality in America focuses on teacher preparation and teacher recruitment. The 2002 NCTAF report, *Unraveling the Teacher Shortage Problem: Teacher Retention Is the Key*, adds another component to the conversation: development of stronger teacher retention strategies.

Teacher Work Rules

Union contracts often stipulate that currently employed teachers meet in the spring semester and decide which classes they want to teach in the coming fall semester. This sounds like a harmless policy, but it's not. In practice, what happens is that the current teachers pick the best classes for the fall—the ones they know will draw motivated students with concerned parents. That work rule creates a huge problem: the hardest classes with the most difficult students to teach are often left for the teachers who are about to be hired.

"Veterans" in Year Two

According to NCTAF, the annual national teacher turnover rate is 16.8 percent (20 percent in urban school districts). Here's a bigger problem: 46 percent of newly hired teachers—the ones given the hardest classes to teach in their first year—leave the profession by year five!

Moreover, according to NCTAF, in 1987–1988 the average teacher in America had 15 years of experience. By the 2007–2008 school year, that dropped, unbelievably, to 1–2 years![17]

Could Your Company Survive?

Think about those numbers. Could your firm compete if (1) nearly 20 percent of your workforce left the company each year, (2) half of your new hires quit within five years, and (3) the average experience of a worker at your company was one to two years?

That's the precise challenge facing American education in the twenty-first century. Teacher turnover, in my opinion, is a significant contributor to the lower scores earned by American students in global and domestic math and science assessment tests.

A Better Question

Unraveling the Teacher Shortage Problem claimed that too much effort is wasted asking the wrong question: How can we find and prepare more teachers? The seminal question our country needs to ask and answer is this: How do we get the good teachers we have recruited, trained, and hired to stay in their jobs?[18]

If We Build It, Will They Stay?

Unraveling the Teacher Shortage Problem put the spotlight on another reason for high teacher turnover rates: outdated nineteenth- and twentieth-century infrastructure. The report noted that "we are sending our children (and teachers) to factory-era schools to prepare them for life and work in the digital age. We continue to expect our children and teachers to succeed in schools that were designed to operate on a 19th-century agricultural schedule, while using teaching and learning approaches suited to the needs of an industrial economy."[19]

Education Doesn't Get the Memo

The report concluded, "The era of solo teaching in our classrooms is over. The industrial age has been surpassed in every major realm of American society except education. Our teachers are walking away from the large, impersonal bureaucracies that our schools have become. If we want professional educators in our schools, we must turn our schools into professional workplaces."[20]

2003: *Building a Nation of Learners*

For America to compete in today's global economy, we must ensure that students develop the skills they need to take them from the classroom to the boardroom.
—BUSINESS HIGHER EDUCATION FORUM, *BUILDING A NATION OF LEARNERS*[21]

The Business Higher Education Forum is the nation's oldest organization of senior business and higher education executives. Its report, *Building a Nation of Learners*, was released in 2003. It repeats the concerns of other organizations and reports by warning that there is a widening skills gap between traditional education training and "the skills actually required to do today's jobs and those of tomorrow."[22]

The New Necessary Skills

While other reports have warned about the importance of a new set of skills required for the workplace, *Building a Nation of Learners* goes a step further: it identifies those skills. Here is its list of nine skills that American workers need to compete in the twenty-first century:

1. Leadership
2. Teamwork
3. Problem solving
4. Time management
5. Self-management
6. Adaptability
7. Analytical thinking
8. Global consciousness
9. Basic communication[23]

The $64,000 Question

The nine skills are indeed important, and the Business Higher Education Forum is to be commended for being the first group to list them. But there is a bigger challenge: How do you teach those skills in the classroom? How can they be incorporated into textbooks and the curriculum?

Much of the curriculum taught in our classrooms and most of the teaching methods employed are more focused on the repetitive workforce skills needed in nineteenth- and twentieth-century America.

How quickly our nation's education system shifts to a skills-based curriculum will determine the strength of America's economic future.

2004: *Sustaining the Nation's Innovation Ecosystem*

> Americans are living off the economic and security benefits of the last three generations' investment in science and education, but we are now consuming capital.
> —PRESIDENT'S COUNCIL OF ADVISORS ON SCIENCE AND TECHNOLOGY, *SUSTAINING THE NATION'S INNOVATION ECOSYSTEM*[24]

Sustaining the Nation's Innovation Ecosystem was published in 2004 by the President's Council of Advisors on Science and Technology, a committee chaired by Robert J. Herbold, the former COO of the Microsoft Corporation.

Doctors Needed!

The report claimed that although our innovation system—which it defined as an interwoven system consisting of human talent, research universities, corporate research and development, venture capital, and government-funded basic research—is the best in the world, "it is threatened by significant changes in the global technical talent pool. For producing blockbuster, industry-generating ideas, we look to exceptional science and engineering talent at the Ph.D. level residing in universities, industry, and the government."[25]

With Apologies to Bill and Steve

Bill Gates and the late Steve Jobs came up with "industry-generating" ideas without a college degree, but most big ideas emanate from extremely bright individuals with PhDs. Focusing on the global talent pool, *Sustaining the Nation's Innovation Ecosystem* compares the number of PhDs earned in the United States with the number earned in Asian universities in a 14-year period. Although the United States led the world when the analysis began in 1987, the report reveals that a significant tipping point in Asia's favor occurred in 1995, when Asian universities produced more natural science and engineering PhDs than American universities did.

The More, the Better

Some might say, "Who cares? The United States has better universities." That is certainly true—for now. Remember, however, one of the key tenets of Yuasa's Phenomenon: the world's center of scientific activity always has more scientific talent. Recall, too, the theory of Harvard professor Niall Fergusson that China's strategy is based on four "mores": "consume more, import more, invest more, and innovate more."[26]

The National Science Foundation numbers presented in *Sustaining the Nation's Innovation Ecosystem* suggest that China, a major location of the Asian universities in the report's analysis, is well on its way to producing *more* PhDs, thereby moving closer to becoming the world's leading scientific center in several decades (see Table 11.10).[27]

Table 11.10 Number of Science and Engineering PhDs

Year	American Universities	Asian Universities
1987	8,238	6,828
1989	8,944	8,117
1991	9,741	8,678
1993	10,033	9,847
1995	10,527	12,303
1997	10,996	15,632
1999	10,586	18,000
2001	10,206	20,000

America went from producing 17 percent more PhDs in science and engineering than Asian universities in 1987 to producing 49 percent less in 2001. Not a good trend.

Sustaining the Nation's Innovation Ecosystem concludes by stating the obvious: "Unless we take action to maintain our global advantage in training the top technical talent and in developing a skilled science and technology workforce, we run the risk of losing our competitive advantage."[28]

2005: *Losing the Competitive Advantage: The Challenge for Science and Technology in America*

The dominance of the U.S. is already over.
—PETER DRUCKER, *FORTUNE* MAGAZINE[29]

One year after the President's Council of Advisors on Science and Technology addressed the risk of America losing its competitive advantage, the American Electronics Association report, *Losing the Competitive Advantage: The Challenge for Science and Technology in America*, said that the loss of competitive advantage by the United States was no longer a risk.

It was reality.

Resting on Our Laurels

Losing the Competitive Advantage explored the competitive challenges the United States faced in 2005 and in many ways continues to face today. The report said the following:

> As the United States takes its leadership for granted, countries around the world have caught on and are catching up. While we begin to close our doors to the best and brightest minds [through more stringent immigration policies], these talented individuals and the intellectual property and jobs they create here are lured elsewhere. As we cut funding for research and development, other countries are investing in research, scientific education, and high technology infrastructure. While we continue to believe know-how and ingenuity are exclusive American brands, dozens of emerging nations are restructuring their economies and challenging our superiority. Americans may be surprised if the next revolutionary technology is produced abroad, but we should not be.[30]

Has America Lost Its Lead in Information Technology?

What do you consider to be the most revolutionary IT products of the last 25 years? I often ask that question to tech execs and get replies like the iPhone, the iPad, and the Netscape browser. Granted, those could be considered revolutionary, but in fact they are impressive *evolutions* of devices and software invented in earlier decades.

For me, the three most revolutionary IT products in the past 25 years, each invented in the early 1990s, are the World Wide Web, the Internet browser, and the Internet search engine.

All of these were invented by non-Americans. Sir Tim Berners-Lee, a British citizen, invented the World Wide Web and the Internet browser, while three Canadian students at Montreal's McGill University invented Archie, the world's first Internet search engine.

Does this suggest that Britain is on the verge of reasserting itself as a Yuasa world scientific center? Certainly not. But it does indicate that the most revolutionary ideas in IT in the past 25 years were *not* invented by Americans.

Don't Be Surprised

Losing the Competitive Advantage, The Challenge for Science and Technology in the United States closes with this warning:

> We are still in the lead, but it is a precarious one. Already other countries are challenging us in key technology arenas. If we don't act now to maintain our competitive edge, we should not be surprised if the next wave of breakthrough technologies is created abroad. U.S. policymakers and industry leaders need to recognize that as we neglect our technology infrastructure—skilled labor, research and development, and a business-friendly environment—many other countries are adopting economic reforms and are directly competing with the United States for talent, innovation, and technology products and services.[31]

2005: *The Knowledge Economy: Is the United States Losing Its Competitive Edge?*

> In today's rapidly evolving competitive world, the United States can no longer take its supremacy for granted.
>
> —TASK FORCE ON THE FUTURE OF AMERICAN INNOVATION, *THE KNOWLEDGE ECONOMY*[32]

The Task Force on American Innovation was an impressive group of 58 technology companies, technology associations, and professional associations like the IBM Corporation, the American Chemical Society, Microsoft, and the Council on Competitiveness. The group's 2005 report, *The Knowledge Economy: Is the United States Losing Its Competitive Edge?* continued the cacophony of reports directly questioning the current and future state of America's competitiveness.

The report claimed that there are five "signs of trouble" for America. Considered in concert, these trouble spots strongly suggest that America is losing its global innovation lead. Here are the five signs of trouble:

1. **Education.** In total volume, the U.S. share of worldwide undergraduate STEM degrees, about 500,000 per year, lags far behind the 1,200,000 STEM degrees annually conferred in Asia and the 850,000 STEM degrees in Europe.
2. **Workforce Benchmark.** Asians are turning away from U.S. universities. From 1994 through 1998 the number of Asian students who decided to pursue their STEM PhD degree dropped from 4,982 to 4,029, while Asians who chose to pursue their STEM PhD in their own countries doubled from 4,983 to 9,942.
3. **Knowledge Creation.** From 1988 to 2001 the United States increased its number of published STEM articles by 13 percent while during the same time the number of western European–published STEM articles rose 59 percent, Japan's total increased by 67 percent, and the [other] countries of East Asia, including China, rose by a staggering 492 percent.
4. **Research and Development.** From 1995 through 2001 the emerging economies of China, South Korea, and Taiwan increased their gross research and development investments by 140 percent. During the same period, the United States increased its base investments by 34 percent. Granted, the Asian country numbers are off a much smaller base, but it doesn't take long to catch up to the United States at that pace.
5. **High-Tech Economy.** The U.S. share of worldwide high-tech exports has been in a 20-year decline. From 1980 until 2001 the U.S. share fell from 31 percent to 18 percent. At the same time, the global share for China, South Korea, and other emerging Asian countries has increased from just 7 percent to 25 percent.[33]

These five signs of trouble scream loud and clear to me that Yuasa's Phenomenon is in play once again.

2005: *The World Is Flat: A Brief History of the Twenty-first Century*

> Today, the U.S., you are the designers, the architects, and the developing countries are the bricklayers for the buildings. But one day I hope we will be the architects.
>
> —XIA DEREN TO THOMAS FRIEDMAN IN *THE WORLD IS FLAT*[34]

Xia Deren, the mayor of Dalian, a major Chinese port city, said this to *New York Times* columnist Thomas Friedman while explaining China's long-term economic plan. It's my favorite quote from Friedman's 2005 best seller, *The World Is Flat: A Brief History of the Twenty-first Century*.

For those who may not have read *The World Is Flat*, the book focuses on 10 "flatteners": macro trends that have leveled the global marketplace. Here's Friedman's list: (1) the collapse of the Berlin Wall, (2) Netscape, (3) workflow software, (4) uploading, (5) outsourcing, (6) offshoring, (7) supply chaining, (8) insourcing, (9) informing, and (10) steroids (wireless).[35]

What I find interesting about Friedman's 2005 list now is how seemingly outdated it is in 2013, a sure indicator of how quickly the world continues to change. I had an idea: Could I come up with a 2013 flattener list representing that pace of change?

I gave it a shot, and with apologies to Friedman, my list of the 10 new flatteners creating global change in 2013 is shown in Table 11.11.

Regardless of whether you think this comparison is silly or valid, it does underscore how much the world has changed in the eight years since *The World Is Flat* was published. The global pace of change continues to

Table 11.11   Global Marketplace Flatteners

	Friedman's 2005 List	Beach's 2013 List
Flattener 1	Collapse of Berlin Wall	The Arab Spring
Flattener 2	Netscape	Twitter
Flattener 3	Workflow software	Cloud computing
Flattener 4	Uploading	4G
Flattener 5	Outsourcing	Global sourcing
Flattener 6	Offshoring	Rural sourcing
Flattener 7	Supply chain	Amazon
Flattener 8	Insourcing	Crowd sourcing
Flattener 9	Informing	Wikipedia
Flattener 10	Steroids	Hyperconnectivity

emphasize how important it is that the American education system undergo a systemic, top-to-bottom restructuring to remain relevant in the twenty-first century.

Such a restructuring has yet to start.

2005: *Rising above the Gathering Storm: Energizing and Employing America for a Brighter Economic Future*

> Every morning in Africa a gazelle wakes up and knows it must outrun the fastest lion or it will be killed. Every morning in Africa a lion wakes up and knows it must outrun the slowest gazelle or it will starve. It doesn't matter whether you are a lion or a gazelle—when the sun comes up, you better be running.
>
> —NORMAN R. AUGUSTINE, RETIRED CHAIRMAN,
> LOCKHEED MARTIN CORPORATION[36]

Rising above the Gathering Storm: Energizing and Employing America for a Brighter Economic Future is perhaps one of the most influential reports included in this book. This 592-page tome was written by the National Academy of Sciences at the request of Senator Lamar Alexander of Tennessee and Senator Jeffrey Bingaman of New Mexico, who cochaired the Senate Committee on Energy and Natural Resources.

The senators asked the National Academy of Sciences, "What are the top 10 actions, in priority order, that federal policymakers could take to enhance the science and technology enterprise so the United States can successfully compete, prosper, and be secure in the global community of the 21st century?"[37]

Setting the Stage

Norman Augustine of the Lockheed Martin Corporation used the gazelle and lion quote above to begin his testimony to the Senate hearing in October 2005 to underscore a main finding: to survive, the United States has to significantly upgrade STEM education at all levels of schooling.

Augustine's observations to the lawmakers became *Rising above the Gathering Storm's* conclusions:

> For the cost of one engineer in the United States, a company can hire 11 in India; 38% of the scientists and engineers in America holding Ph.D. degrees were born abroad; the United States is now a net importer of high technology products; about two-thirds of students

studying chemistry and physics in U.S. high schools are taught by teachers with no major or certificate in the subject; 50% of middle school and high school are taught by teachers who did not major in math, and many students are taught math by physical education majors; in 2001 the United States spent more on tort litigation and related costs than on research and development; in 2003 only three American companies ranked among the top 10 recipients of patents granted by the U.S. Patent Office.[38]

The list gets your attention, doesn't it?

Flying above the Gathering Storm

Although the Senate hearing requested 10 recommendations, the blue-ribbon panel of 20 government, industry, and academic leaders distilled their recommendations to a list of four. Three of those moved quickly from report recommendations to becoming prominent provisions in the America COMPETES Authorization Act of 2007 (COMPETES stands for Creating Opportunities to Meaningfully Promote Excellence in Technology Education and Science).

Three Big Ideas

These are the three recommendations made by *Rising above the Gathering Storm* that were included in the America COMPETES Act:

1. Increase America's talent pool by vastly improving K–12 science and mathematics education by (a) annually recruiting 10,000 science and math teachers who would earn a bachelor's degree in physical science, engineering, or mathematics, (b) strengthening the skills of 250,000 math and science teachers through training and education programs, (c) enlarging the pipeline of students who are prepared to enter college and graduate by increasing the number of students who pass science and/or math advanced placement courses.

2. Sustain and strengthen the nation's traditional commitment to long-term research by (a) increasing the federal investment in long-term basic research by 10 percent over the next seven years, (b) providing new research grants of $500,000 each to 200 of the nation's most outstanding early-career researchers, (c) allocating at least 8 percent of the budgets of federal research agencies to discretionary funding, (d) creating in the Department of Energy an organization like

[the Defense Department's] DARPA (which invented the Internet) and call it the Advanced Research Projects Agency–Energy and (e) instituting a Presidential Innovation Award to stimulate scientific and engineering advances in the national interest.

3. Make the United States the most attractive setting in which to study and perform research so that we can develop, recruit, and retain the best students, scientists, and engineers from within the United States and throughout the world by (a) increasing the number and proportion of U.S. citizens who earn bachelors' degrees in the physical sciences, life sciences, engineering, and mathematics by providing 25,000 four-year scholarships to U.S. citizens attending U.S. universities, (b) increasing the number of U.S. citizens pursuing graduate study in areas of national need by funding annually 5,000 graduate fellowships, (c) providing a federal tax credit to encourage employers to make continuing education available to practicing scientists and engineers, (d) continuing to improve visa processing for international students and scholars, (e) providing a one-year automatic visa extension to international students who receive their doctorates in STEM to remain in the United States.[39]

Show Me the Money

The problem with many Congressional bills that authorize something new is that although they look good on paper, they have to be followed by appropriations bills that fund them.

The 2007 America COMPETES Act was signed into law by President George W. Bush on August 9, 2007, with provisions like "Teachers for a Competitive Tomorrow," which recommended an increase in the number of professionals with STEM teacher certification. The act also called for the creation of a grant program to facilitate a master teacher program in which outstanding STEM teachers could influence other teachers.

The Great Recession of 2008 began five months after the bill was signed, however, and other fiscal priorities dominated the agendas of federal politicians for the next two years. The American COMPETES Act was finally funded, but at a level that was only half of the bill's recommended level.

The America COMPETES Act was reauthorized in January 2010 and signed into law on December 21, 2010, right after the 2010 Congressional midterm elections in which the Republicans took control of the House of Representatives. Even in the highly partisan political atmosphere of Washington, the broad provisions of the America COMPETES Act continue to get bipartisan support.

Is the America COMPETES Act Our Country's National Technology Strategy?

Boston University professor Jeffrey L. Furman, who is also a member of the National Bureau of Economic Research (NBER), the group of economists that officially declares when a U.S. recession begins and ends, wrote an analysis of the America COMPETES Act in 2012 in which he said the following:

> The America COMPETES Reauthorization Act of 2010 was one of the prominent bipartisan legislative achievements of the past decade and was seen as having the potential to be the most notable science and innovation policy initiative of the new millennium. The aims of the COMPETES Act were to substantially increase the extent of federal funding for physical science and engineering research in the United States and to improve the country's research infrastructure and STEM capabilities in these areas.[40]

Homework

The National Academy of Sciences and the authors of *Rising above the Gathering Storm*, particularly Norm Augustine, must be commended for their cogent recommendations to the Senate that framed the first America COMPETES Act of 2007.

The 2010 reauthorization of the America COMPETES Act is our nation's best example of a national technology plan and outlines well the path our country needs to take to compete better in STEM.

Watching grass grow is more exciting than reading congressional bills; nevertheless, I recommend that every reader do a computer search for "America COMPETES Act 2010" and read it. It is quite a thorough document. And since the provisions in the America COMPETES Act have bipartisan support, the recommendations you're reading about here actually have a chance of becoming widely funded in the future.

2005: The National Assessment of Educational Progress

Time to check in again with the National Assessment of Educational Progress results and review the 2005 Nation's Report Card for eighth-grade students three years after the mandatory testing provisions of the No Child Left Behind Act became the law of the land.

On the positive side, eighth graders improved their scores in math compared to the 2000 test results (see Table 11.12). The results for science, however, continued to show significant erosion (see Table 11.13).[41]

Table 11.12   Comparison of Math Scores, 2000 and 2005

	Advanced	Proficient	Basic	Below Basic
2000	5%	22%	38%	34%
2005	5%	30%	44%	21%

Table 11.13   Comparison of Science Scores, 2000 and 2005

	Advanced	Proficient	Basic	Below Basic
2000	4%	26%	36%	34%
2005	3%	24%	30%	43%

Let Me "C"

Keep in mind that *advanced* can generally be equated with an A grade, *proficient* with a B, *basic* with a C, and *below basic* with a D or worse. Reviewed that way, these results are not exactly encouraging: 65 percent of American eighth graders earned a C or worse in math, and 73 percent earned a C or worse in science.

2006: *Teachers and the Uncertain American Future*

"Teaching is not a lost art, but the regard for it is a lost tradition."
—EDUCATIONAL TESTING SERVICE, *TEACHERS AND THE UNCERTAIN AMERICAN FUTURE*[42]

With this report, the Educational Testing Service (ETS), the group that administers the SAT, weighed in with six compelling recommendations to improve the profession of teaching in America.

Here Today, Gone Tomorrow

The report stated bluntly that America's future is "uncertain" because of a "failure of American vision and leadership" and that a problem of "epic proportion looms on the horizon—a problem that has yet to register fully with the nation."

According to the ETS, that problem is the high turnover rate for teachers, in which "46% of the new teachers who enter elementary and secondary schools leave the classroom within five years" and "nearly half of the current teachers may be looking for retirement."[43]

Six Bold Recommendations

The report offered six solutions to the crisis:

1. Raise teacher salaries immediately by 15–20 percent and up to 50 percent in the foreseeable future.
2. Implement three levels of teachers (Beginning, Professional, and Instructional).
3. Create multiple pathways into teaching.
4. Mount intense recruitment programs targeted at minority students to become teachers.
5. Increase by 50 percent the number of young people entering science, mathematics, and engineering professions.
6. Create a national "Teachers' Trust" to fund the prior five initiatives.[44]

I support the idea of raising teacher salaries with the goal of attracting and retaining outstanding teachers, not just giving all teachers a raise. I also like the recommendation of hiring more "minority" teachers, with the stipulation that that include male teachers. Just as the IT industry is heavily dominated by men, the teaching profession is heavily dominated by women.

But my favorite recommendation by far is the creation of "multiple pathways" into the teaching profession. Teacher recruitment is so one-dimensional, achieved mostly by hiring college graduates. America needs more men and women with STEM degrees who are currently employed in the workforce to step up and say, "I want to change careers and teach." I am convinced there are tens of thousands of IT workers who would be willing to do so.

More Trust

The plan to fund a Teachers' Trust with more state and local taxes is not feasible. But I am enamored with the idea of a trust, and in the Epilogue I will offer a different approach to funding one.

2006: *The Quiet Crisis: Falling Short in Producing American Scientific and Technical Talent*

"Who Will Do Science?"[45] This question was framed as the title of a 1994 book edited by Willie Pearson Jr. and Alan Fechter. It was also addressed in *The Quiet Crisis: Falling Short in Producing American Scientific and Technical Talent*, a series of provocative articles and speeches written by Shirley Ann

Jackson, the president of Rensselaer Polytechnic Institute in Troy, New York. Here's a 2006 excerpt in which Jackson explains what she means by the "quiet crisis":

> There is a quiet crisis building in the United States—a crisis that could jeopardize the nation's preeminence and well-being. The crisis has been mounting gradually, but inexorably, over several decades. If permitted to continue unmitigated, it could reverse the global leadership Americans currently enjoy.
>
> The crisis stems from the gap between the nation's growing need for scientists, engineers, and other technically skilled workers and its production of them. As the generation educated in the 1950s and 1960s prepares to retire, our colleges and universities are not graduating enough scientific and technical talent to step into research laboratories, software and other design centers, refineries, defense installations, science policy offices, manufacturing shop floors, and high-tech start-ups. This gap represents a shortfall in our national scientific and technical capabilities.
>
> The need to make the nation safer from emerging terrorist threats that endanger the nation's people, infrastructure, economy, health, and environment makes this gap all the more critical and the need for action all the more urgent.
>
> We ignore this gap at our peril. Closing it will require a national commitment to develop more of the talent of all our citizens.[46]

Jackson remains one of America's most prominent spokespeople on the STEM issue. Visit the web site and read all of her essays on the STEM challenges facing America.

2007: *We Are Still Losing Our Competitive Advantage: Now Is the Time to Act*

> Two years ago the AEA called the United States the proverbial frog in the pot of water, oblivious to the slowly rising temperature of a world catching up to us.
>
> —AMERICAN ELECTRONICS ASSOCIATION,
> *WE ARE STILL LOSING OUR COMPETITIVE ADVANTAGE*[47]

Only two years after issuing its first report, the American Electronics Association issued a second report on America's competitive future, *We Are*

Still Losing Our Competitive Advantage: Now Is the Time to Act—the operative word being *still*. Why the need for another report so soon?

Congressional Gridlock Threatens America's Future

The American Electronics Association was not pleased with Washington political gridlock. *We Are Still Losing Our Competitive Advantage* claimed that "much has changed in the last two years. America's political leaders have become aware that more and more countries, companies, universities, and individuals around the world are trying to out-compete us. And yet we have not moved forward."[48]

Undeterred by the lack of action in Washington, the association presents six additional recommendations in its new report. These recommendations are grouped into two tiers.

Tier One Recommendations

1. Champion dramatic improvements in the U.S. educational system by (a) sustaining, strengthening, and reauthorizing the No Child Left Behind Act, (b) promoting undergraduate and graduate level STEM education, and (c) creating a human capital investment tax credit to promote continuous education.
2. Support and increase research and development by (a) increasing federal funding for basic research, and (b) strengthening the research and development tax credit and making it permanent.
3. Enact high-skill visa reform by (a) lowering the barriers for high-skilled individuals to receive temporary work visas (H-1B visas), and (b) giving green cards (which give permanent-resident status to foreign nationals) to all U.S. educated master and doctoral students.

Tier Two Recommendations

1. Create a more business-friendly environment by (a) reducing the onerous tax burden on small and medium-size business created by Sarbanes-Oxley's 404 compliance code, (b) funding the U.S. Patent Office to reduce lag times, and (c) addressing the rising costs of health care.
2. Engage proactively in the global trade system by (a) advancing fair trade policies, (b) renewing the president's trade promotion authority, and (c) enforcing stronger intellectual property protection rules worldwide.
3. Promote broadband by (a) providing industry the incentives to deploy broadband infrastructures, and (b) ensuring access to broadband for every American by 2012.[49]

Still Waiting after All These Years

What impressed me most about this report was that the association leadership deemed it necessary to issue *two warnings within 24 months on the identical topic*: the future competitiveness of the United States.

For the American Electronics Association, an association representing 3,000 technology companies, the water in the America's competitive pot was approaching 212 degrees. The group clearly questioned America's ability to jump out of that threatening pot.

2007: *How the World's Best-Performing School Systems Come Out on Top*

> Success will go to those individuals and countries which are swift to adapt, slow to complain, and open to change. The task for government will be to ensure that countries rise to this challenge.[50]
>
> —ANDREAS SCHLEICHER, DIRECTORATE FOR EDUCATION, **OECD**

With more than a decade of results from international math and science assessment tests available for review, McKinsey & Company, a global management consulting firm, decided to study the quantitative and qualitative factors that high-performing schools around the world have in common. The result was the 2007 report *How the World's Best-Performing School Systems Come Out on Top*.

Don't Show Me the Money

McKinsey & Company quickly determined that the amount of money invested in public education by a country had no direct effect on the performance of students in international science and math assessment tests. The report claims, for example, that "Singaporean students score at the top of TIMSS assessment tests despite the fact that Singapore spends less on each student in primary education than almost any other developed country."

Compared to the fiscally frugal Singapore education system, "between 1980 and 2005, public spending per student increased by 73% in the United States (allowing for inflation), the U.S. employed more teachers, the student-to-teacher ratio fell by 18%, and class sizes were the smallest they had ever been . . . yet student outcomes stayed almost the same."[51]

The Holy Grail

McKinsey & Company then identified three factors that did have a direct effect on improving public education around the world.

1. Get the Right People to Become Teachers

The report was adamant that "the quality of an education system cannot exceed the quality of its teachers." Offered as an example was a research study from Tennessee that revealed that "if two average eight-year-old students were given different teachers—one of them a high performer and the other a low performer—the student performance would diverge by more than 50 percentile points within three years." The report reviewed other case studies from Boston and cities in England and came to this blunt conclusion: "Primary students who are placed with low-performing teachers for several years in a row suffer an educational loss which is largely irreversible."

What is a "quality" or "high-performing" teacher? McKinsey & Company claimed that "in the United States, studies show that a teacher's level of literacy, measured by vocabulary and other standardized tests, affects student achievement more than any other measurable teacher attribute."[52]

Where do you find "quality" or "high-performing" teachers?

McKinsey & Company discovered a remarkable difference in how and where different countries recruit new teachers. South Korea recruits teachers solely from the top 5 percent of college graduates, particularly for elementary grade levels. Finland focuses on the top 10 percent of graduates, and Singapore and Hong Kong widen the scope to the top 30 percent. The United States, however, primarily recruits teachers from the bottom third of college graduates.

2. Develop Teachers into Effective Instructors

How the World's Best-Performing School Systems Come Out on Top quotes a Boston educator who, paraphrasing the famous quote about real estate, says that the three most important "pillars of reform are professional development, professional development, and professional development."

The report listed three keys to professional development:

(1) teachers need to become aware of specific weaknesses in their own practice, (2) teachers need to gain understanding of specific best practices, and (3) teachers need to be motivated to make the necessary improvements.[53]

The report also emphasized the importance of teachers being able to learn from one another on a regular basis, not just when a supervisor enters their classroom for a 20-minute observation. In the Epilogue I will make a somewhat radical suggestion on how America can implement a bold peer-based teacher learning program.

3. Have Clear Goals, High Standards, and Transparent Results

The third characteristic of high-performing national education systems was "that every child, rather than just some children, have access to excellent instruction. All of the top-performing and rapidly improving systems have curriculum standards which set clear and high expectations for what students should achieve." The report strongly underscores the importance of testing, of benchmarking schools in terms of a set of key performance indicators and annual external reviews. The transparency of success or failure in student tests and teacher reviews was considered important, and the report shares an example from New Zealand in which a policy maker said, "We make everything public; it creates tension in the system—transparency over the problems—and that drives improvement."[54]

The Puzzle Comes Together

When I read *How the World's Best-Performing School Systems Come Out on Top*, key pieces of the puzzle on how to fix American education started to fit in place. If you combine (1) the results of the McKinsey & Company analysis of where to find great teachers, (2) the results of the 2002 NCTAF report on teacher retention (discussed earlier in this chapter), and (3) the public release of student test scores and teacher reviews, you can actually can see a solution forming: hire great teachers, work hard to keep them on the payroll, and publicly release education results, not unlike quarterly financial business reports.

Release the Scores

Some people will express concerns about privacy issues on that last point, and I can understand that. But people who are against the public release of student and teacher review data must consider this: the American public, and the $583 billion it spends each year on public education, are the "shareholders" (*stakeholders*, in education jargon), and they have every right to demand to know what the "score" is.

I strongly believe that the public release of student test scores is exactly the kind of peer pressure American students need.

Of course, obstacles like teacher unions and salaries remain. But if the United States agrees that the solution to swimming out of the sea of mediocrity

is to get the best and smartest teachers in front of our nation's classrooms—
and make sure they stay there longer than two years—we can make progress
on the other issues.

2007: *Into the Eye of the Storm: Assessing the Evidence on Science and Engineering Education, Quality, and Workforce Demand*

> Despite this nearly universal support for upgrading science and math
> education, our review of the data leads us to conclude that it is not as
> dysfunctional as believed.
> —B. LINDSAY LOWELL AND HAROLD SALZMAN, *INTO THE EYE OF THE STORM*[55]

Oh, no, another storm metaphor!

But this report, *Into the Eye of the Storm: Assessing the Evidence on Science
and Engineering Education Quality and Workforce Demand*, reveals that not everybody was buying into the dire predictions. B. Lindsay Lowell of Georgetown University and Hal Salzman of the Urban Institute harbored a different
perspective.

More Than Enough Smarts

Lowell and Salzman studied data on the global science and engineering
workforce and claimed that "today's American high school students actually test as well or better than students two decades ago. Furthermore,
today's students take more science and math classes, and a large number
of students with strong science and math backgrounds graduate from
U.S. high schools and start college in science and engineering fields of
study."

Into the Eye of the Storm examined the issue of whether America's higher
education institutions are producing enough STEM experts and concludes
that "the pool of qualified graduates is several times larger than the number
of annual job openings. Our analysis at the aggregate level does not find a
shortage of potential students or workers."[56]

Who Cares about Comparisons to Singapore and Norway?

Lowell and Salzman refuted the poor showing of U.S. students in international assessment tests. They wrote, "It is difficult to conclude that the major
economic 'threats' to the United States are related to the performance of U.S.

students as compared to students in other countries. Our major economic competitors, particularly emerging nation behemoths, are not among the top test scoring nations."[57]

Major Industrialized Nations Outperform the United States, Too

When I read that comment by Lowell and Salzman, I decided to do my own research. I gathered data from 17 years' worth of Trends in International Mathematics and Science Study (TIMSS) math and science assessment scores for eighth graders from the G7 nations (the United States, Canada, France, Japan, Italy, Britain, and Germany) plus Russia and China.

I was curious: Did Lowell and Salzman have a point? Were critics making too much fuss in America over the United States constantly trailing emerging countries in these global math and science assessment tests? I also wondered how America was performing compared to larger, industrialized countries.

My Method of Madness

Here's what I did. I added up the raw math and science scores for eighth graders in the TIMSS examinations and then divided that by the number of TIMSS tests that each country participated in since 1995.

Here's what I found.

Lowell and Salzman are wrong. Not only does America have to worry about *emerging* countries, we also should be concerned, and probably even more so, with the science and math skills of industrialized countries, which Lowell and Salzman label America's "major economic competitors."

In my comparison with the G7 plus Russia and China, the United States places near the bottom third of the list. Japan, China, Canada, Russia, and Britain all place ahead of America. Table 11.14 shows these results.[58]

The Wrong Questions

Lowell and Salzman unfortunately conclude their report with the wrong questions: "Should U.S. policy be driven by the test score performance of students in Flemish Belgium, Latvian-speaking Latvia, or even Singapore? How will these countries find the capital and the numbers of workers needed to 'steal' any major portion of a U.S. industry?"[59]

Maybe they will and maybe they won't. But U.S. policy should be driven by how we compare to our major economic competitors. Such a comparison right now does not favor the future economic prospects of the United States. Powerful, relevant countries do have the talent to "steal [a] major portion of a U.S. industry."

Table 11.14   U.S. Students Also Lag behind Developed Nations

Country	17-Year Average in TIMSS Math and Science
1. Japan	565
2. China	554
3. Canada	519
4. Russia	518
5. Britain	517
6. United States	**509**
7. Germany	506
8. France	496
9. Italy	486

Source: TIMSS 1995–2007

2007: *Tough Choices or Tough Times*

The core problem is that our education and training systems were built for another era. We can get where we must go only by changing the system itself.

—COMMISSION ON THE SKILLS OF THE AMERICAN WORKFORCE,
TOUGH CHOICES OR TOUGH TIMES[60]

Tough Choices or Tough Times was the follow-up report to the 1990 *America's Choice: High Skills or Low Wages!* report by the Commission on the Skills of the American Workforce, appointed by the National Center on Education and the Economy. The 1990 report was one of the first to identify the shift to a service economy and the new skills required to compete in the global workforce.

Mind the Gap

Tough Choices or Tough Times warned the following:

If we continue on our current course, and the number of nations outpacing us in the education race continues to grow at its current rate, the American standard of living will steadily fall relative to those nations, rich and poor, that are doing a better job. If the gap gets to a certain—unknowable—point, the world's investors will conclude they can get a greater return on their funds elsewhere, and it will be almost impossible to reverse course. Although it is possible to construct a scenario for improving our standard of living, the clear and present danger is it will fall for most Americans.[61]

What's the Problem? Change the Schools

The Commission on the Skills of the American Workforce stated that public education is the "core" problem facing our nation. The American education system, the report said, must be changed, and *Tough Choices or Tough Times* recommended three ways to change it.

1. Change Teacher Recruitment Practices

Similar to the findings of the McKinsey & Company report reviewed earlier, *Tough Choices or Tough Times* recommended the following:

> We recruit a disproportionate share of our teachers from among the less able of high school students who go to college. Many of our teachers are superb. But we have for a long time gotten better teachers than we deserved because of the limited opportunities for women and minorities in our workforce. Those opportunities are far wider now, and we are left with the reality that we are now recruiting more of our teachers from the bottom third of the high school students going to college than is wise. We must recruit many more from the top third.[62]

2. Front-Load Teacher Compensation Packages

How teachers are compensated requires radical change. "Our teacher compensation system is designed to reward time in service, rather than attract the best and brightest of our college students" to become teachers. The report continues:

> The shape of teacher compensation is backloaded, in the sense that it is weak on cash compensation, especially up front, and heavy on pensions and health benefits for the retired teacher. This makes no sense if what we are after is to attract young people who are thinking most about how they are going to get the cash they need to enjoy themselves, buy a home, support a family, and pay for college for their children. The first step in our plan is to make retirement benefits comparable to those of the better firms in the private sector and use the money that is saved from this measure to increase teachers' cash compensation.[63]

3. Create a Common School System

Tough Choices or Tough Times then introduces a radical restructuring of the twenty-first-century American education system from kindergarten to college.

Here's how it would work. Preschool and kindergarten programs would remain as is. A common school, encompassing grades 1–10, would replace primary school, middle school, and the first two years of high school. At the completion of the tenth grade, all students would be required to take a state board qualifying examination. Based on the results of that test, the students would be given the choice of going to either an upper secondary academic program (two years, offering mostly advanced placement courses), or a regional vocational school, community college, or technical college (two to three years). At the completion of either of those two options, the students would enroll in a four-year baccalaureate program or take the technical or state board exam required to enter a profession.

Setting an Agenda

For years, critics have railed that the American public school system is steeped in ways of the nineteenth-century manufacturing-based economy. The ideas presented in *Tough Choices or Tough Times* set an agenda to radically restructure the American education system.

2007: *The Role of Education Quality in Economic Growth*

"There is strong evidence that the cognitive skills of a population, rather than mere school enrollment, are powerfully related to economic growth." So claimed the World Bank in a 2007 report entitled *The Role of Education Quality in Economic Growth*, which then asked this provocative question: "Loosely speaking, is it a few 'rocket scientists' at the very top who spur economic growth, or is it 'education for all' that lays a broad base at the lower parts of the distribution?"[64]

Smarts Count

Eric A. Hanushek and Ludger Woessmann, the authors of the report, wrote the following:

> Access to education is one of the highest priorities on the development agenda. Significant results have already been achieved in school enrollment. Yet care must be taken that the need for simple,

measurable goals does not lead to ignoring the fact that it ultimately is the degree to which schooling fosters cognitive skills and facilitates the acquisition of professional skills that matters for development. Differences in learning achievement matter more in explaining productivity growth than differences in the average number of years of schooling or enrollment rates.[65]

The Four Pillars of Quality Education

Hanushek and Woessmann identified four components of education quality:

1. Effective teacher certification.
2. Public disclosure of the educational achievements of schools and teachers.
3. Local school control by parent associations.
4. Accountability of teachers and head teachers.

Moreover, in a stunning proclamation, particularly for a country like the United States, which significantly outspends the rest of the world with its $583 billion annual budget for public education, the report declared, "Simply increasing educational spending does not ensure improved student outcomes."[66]

The Quality of America's Education System

I decided to take Hanushek and Woessmann's four pillars and do a personal evaluation to determine whether our nation's education system meets their criteria. Here's how the U.S. system measures up:

1. **Certification.** The United States does have a broad-based teacher certification program. It is very difficult to secure a public school teaching job without being certified. In my opinion, however, America is certifying the wrong issue—namely, that an individual knows how to teach. Far more important, particularly for the subjects of science and math, is a system that mandates that teachers of those subjects have certifiable undergraduate, or higher, degrees in their STEM subjects. To its credit, the America COMPETES Act, discussed earlier, has attempted to jump-start this with its recommendation to add 10,000 new math and science teachers a year to the nation's teacher ranks, but the recommendation has yet to be widely implemented. Even though nearly all U.S. teachers must have some sort of certification, the majority of science and math teachers continue to teach those subjects out of field.

2. **Public disclosure of educational achievements.** The No Child Left Behind Act started the thorny challenge of publicly disclosing which schools are passing and which schools are failing, but many believe that the data are flawed. Their point is this: How can you determine if an entire school is failing? What if there are pockets, perhaps broad ones, of excellence, in an otherwise failing school?

   There is no effort in the United States to publicly disclose, for instance, the NAEP scores, the only publicly funded test in the country. Can you imagine the public outcry if the U.S. Department of Education made a decision to release to the public the NAEP scores of every twelfth grader in America?

3. **School control.** This one is simple. Parents and taxpayers underwrite public schools. Teacher unions and administrators control them.

4. **Teacher accountability.** This is a thorny issue. What do you measure and how do you measure it? One idea I have harbored for a long time is to allow parents and students, the "customers" of teachers, to vote each year on the performance of the teachers. The New York City public school system almost introduced a highly controversial public teacher rating system. The system, which produced a document called *Teacher Data Reports*, was intended for private use. The media, most notably the *Wall Street Journal* and the *New York Times*, sued for access to the evaluations under the provisions of the Freedom of Information Act and obtained the results. *Teacher Data Reports* is based on a "value-added analysis that calculates a teacher's effectiveness in improving student performance on standardized tests—based on past test scores. The forecasted figure is compared to the student's actual scores, and the difference is considered the value added, or subtracted, by the teachers."[67]

   Come on! That's not the way to evaluate great teaching! I guarantee that every reader of this book, if asked, could immediately identify a "great teacher" in his or her life. And I guarantee that the reason has nothing in common with the New York City criteria, a system that is inherently biased because of teacher union work rules that allow currently employed teachers to select in the spring the most favorable classes to teach in the fall.

   We need a better way to evaluate great teaching, and in this age of social media, we need to create something akin to Trip Advisor, the Internet-based social media site that rates hotels and airlines, for teacher evaluation. Who cares about the stodgy school administrators and teacher unions? Students and their parents are the "customers" of teachers. They know best who the great teachers are all across

America. It is time we leverage social media to rate millions of American teachers.

I will come back to this idea in the Epilogue. The No Child Left Behind approach to rating schools doesn't work, and there is currently no widespread adoption of publicly rating teachers.

The United States has a long way to go until we have a quality education system that embraces transparent, publicly released results.

More important, if the World Bank's forecast is correct that a country's future economic growth is intrinsically tied to the quality of that country's education system, the future economic growth of the United States faces a long, uphill battle.

2008: *Foundations for Success: The Final Report of the National Mathematics Advisory Panel*

The concerns of national policy relating to mathematics education go far beyond those in our society who will become scientists or engineers. The national workforce of future years will have to handle quantitative concepts more fully and deftly than at present.

—NATIONAL MATHEMATICS ADVISORY PANEL, *FOUNDATIONS FOR SUCCESS: THE FINAL REPORT OF THE NATIONAL MATHEMATICS ADVISORY PANEL*[68]

Foundations for Success: The Final Report of the National Mathematics Advisory Panel, was underwritten by the Department of Education. It emphasized that a nation skilled in math is not a future option, it's a necessity.

What Must Be Taught?

Foundations for Success claimed there are three key elements in improving the math skills of American students. First, the prekindergarten through eighth-grade math curriculum has to be streamlined and emphasize a well-defined set of the most critical topics, particularly in the early grades. (This issue was addressed in Chapter 5, where we noted that Bill Gates advocated the same idea as he criticized 300-page math and science textbooks as "mind-blowing.")

Second, the report encouraged the adoption of "rigorous" evaluation initiatives for "attracting and appropriately preparing prospective teachers and for evaluating and retaining effective teachers."

Third, *Foundations for Success* said educators in America need to learn "what is clearly known from rigorous research about how children learn mathematics."[69]

The Devil in the Details

Let's review each recommendation, starting with the suggestion to streamline the curriculum. My first job was with Macmillan Publishing Company, where I worked in the marketing department for *Teacher* magazine, a publication for K–8 teachers. That experience taught me how difficult the first recommendation of *Foundations for Success* would be. K–12 textbook publishers are an insular industry. They focus on convincing state textbook adoption boards, powerful groups that can make or break a textbook publisher's business, that their textbook program is different from that of their competitors. Could they migrate to a "streamlined" model? Maybe. But the reason your kids have 300-page math and science textbooks by middle school is the need of a textbook publisher to differentiate itself from all the others.

The second recommendation of *Foundations for Success* is for "rigorous evaluation initiatives." Unfortunately, this idea is dead on arrival because of the uncompetitive starting-salaries offered by most of the 15,000 school districts in America. Offering salaries of $30,000 or so, school districts cannot possibly add "rigorous evaluation initiatives" to a teacher's workload. No one would apply. We certainly need more rigorous evaluation initiatives, but they must go hand in hand with a higher starting salary.

The National Mathematics Advisory Panel's final recommendation, that educators should learn more about how students learn mathematics, sounds good and makes sense. The question is this: How does the education system incentivize teachers to do this?

2008: "Lessons from 40 Years of Education Reform"

Louis Gerstner, the former chairman and CEO of IBM and a business executive who has always had a passion for improving public education in the United States, wrote an op-ed on the state of public education that appeared in the *Wall Street Journal* on December 1, 2008, just weeks after Barack Obama was elected president. Some believe this opinion piece was directed at the president-elect. Here's what Gerstner said:

> While the economic news has most Americans in a state of near depression, hope abounds today that the country may use the current economic crisis as leverage to address some longstanding problems. Nowhere is that prospect for progress more worthy than the crisis in our public education system. So, from someone who realized rather glumly that he has been working at school reform for 40 years, here is a prescription for leadership for the Obama administration.

We must start with the recognition that, despite decade after decade of reform efforts, our public K–12 schools have not improved. We can point to individual schools and some entire districts that have advanced, but the system as a whole is still failing. High school and college graduation rates, test scores, and the number of graduates majoring in science and engineering all are flat or down over the past two decades. Disappointingly, the relative performance of our students has suffered compared to those of other nations. As a former CEO, I am worried about what this will mean for our future workforce.

It is most crucial for our political leaders to ask why we are at this point—why after millions of pages, in thousands of reports, from hundreds of commissions and task forces, financed by billions of dollars, have we failed to achieve any significant progress?

Answering this question correctly is the key to finally remaking our public schools.

This is a complex problem, but countless experiments and analyses have clearly indicated we need to do four straightforward things to bring fundamental changes to K–12 education: (1) set high academic standards for all of our kids, supported by a rigorous curriculum, (2) greatly improve the quality of teaching in our classrooms, supported by substantially higher compensation for our best teachers, (3) measure student and teacher performance on a systematic basis, supported by tests and assessments, and (4) increase "time on task" for all students; this means more time in school each day, and a longer school year. Everything else either doesn't matter (e.g., smaller class size) or is supportive of these four steps (e.g., vastly improved schools of education).

Lack of effort is not the cause of our 30-year inability to solve our education problem. Not only have we had all those thousands of studies and task forces, but we have seen many courageous and talented individuals pushing hard to move the system. Leaders like Joel Klein (New York City), Michelle Rhee (Washington, DC), and Paul Vallas (New Orleans) have challenged the system, and elected officials from both sides of the political spectrum have also fought valiantly for change. So where does that leave us? If the problem isn't "what we do," nor is it a failure of commitment, what else is stopping us?

I believe the problem lies with the structure and governance of our public schools. We have over 15,000 school districts in America; each of them, in its own way, is involved in standards, curriculum, teacher selection, classroom rules, and so on. This unbelievably unwieldy structure is incapable of executing a program of fundamental change. While we have islands of excellence as a result of great reform programs, we continually fail to scale up systemic change.

I recommend the following: (1) abolish all school districts save 70 (50 states; 20 largest cities). Some states may choose to leave some of the rest as community service organizations, but they would have no direct involvement in the critical task of establishing standards, selecting teachers, and developing curricula; (2) establish a set of national standards for a core curriculum and start with four subjects: reading, math, science, and social studies; (3) establish a National Skills Day on which every third, sixth, ninth, and twelfth grader would be tested against national standards and those results would be published nationwide for every school in America; (4) establish national standards for teacher certification and require regular reevaluations of teacher skills and increase teacher compensation to permit the best teachers (measured by advances in student learning) to earn well in excess of $100,000 per year; and (5) extend the school day and the school year to effectively add 20 more days of schooling for all K–12 students.[70]

2009: *Rising Tigers, Sleeping Giant: Asian Nations Set to Dominate the Clean Energy Race by Out-Investing the United States*

Asia's rising clean technology tigers—China, Japan, and South Korea—will out-invest the United States by three to one over the next five years. With each passing day it seems more likely that future American clean energy workers will be employed by a foreign firm.

—ROB ATKINSON AND MICHAEL SHELLENBERGER,
RISING TIGERS, SLEEPING GIANT[71]

Clean energy is certainly one of the key industries of the future, and it requires a workforce with strong math and science skills. The Breakthrough Institute and the Information Technology and Innovation Foundation produced this 2009 report, which claims to be "a comprehensive comparison of public investments by the United States and key Asian competitors in core clean energy technologies, including solar, wind, nuclear power, carbon capture, advanced vehicles and batteries, and high-speed rail."[72]

Makes Cents

A key premise, and a concern for the United States, of *Rising Tigers, Sleeping Giant* is that the amount of public sector investment in clean energy by the governments of China, Japan, and South Korea is three times that of the United States. According to Rob Atkinson and Michael Shellenberger, the authors of *Rising Tigers, Sleeping Giant*, it means that future private sector investment dollars will most likely flow to where public sector investments are being made.

And where the dollars go, so go the jobs.

United States to Be Net Importer of Clean Energy Technology in the Future

The key consequence of America's uncompetitive clean energy public investment strategy is this: in the future, America will "import the overwhelming majority of clean energy technologies it deploys." The report forecasts that "over the next five years the United States is poised to invest $172 billion (in clean energy projects), which compares to investments of $397 billion in China alone. While some U.S. firms (large multinationals) will benefit from the establishment of joint ventures overseas, the jobs, tax revenues, and other benefits of clean energy growth will overwhelmingly accrue to Asia's clean tech tigers."[73]

Don't Throw Out Those Currency Converters!

Clean energy is one of the most important and innovative industries for America's future. Americans with strong skills in STEM could find good-paying future jobs in them in the United States. But with America's political leaders being more focused on domestic demand-side legislation, like cap and trade, instead of moving aggressively to stimulate a new domestic clean energy manufacturing industry, the report concludes, "with each passing day it seems more likely that those future American clean energy workers will be employed by a foreign firm."[74]

2009: The CIO Executive Council's Youth and Technology Careers Survey

The CIO Executive Council, launched in 2004, is part of IDG Enterprise, the company that also publishes *CIO* magazine. The council is composed of hundreds of the world's leading CIOs. Through peer interaction, collaboration, and research, the CIO Executive Council members tackle the most pressing issues in technology.

One of those issues is a concern about how America's teenagers perceive careers in IT. In 2009, the council's Youth in IT peer group commissioned IDG Research to survey American teenagers about those perceptions. Some of the findings of the Youth and Technology Careers Survey will shock you.

First, here's a look at the ages of the more than 300 respondents: 64 percent were 13–17, 24 percent were 18–21, 8 percent were 21 or older, and 4 percent were under 13.

When asked, "What does the professional term *IT* mean?," 59 percent answered correctly, 33 percent had no idea, and 0.3% said it meant "industrial tracking." Things got worse when they were asked what *CIO* means. The largest group of respondents, 38 percent, had no idea; 22 percent answered correctly, and 19 percent said it meant "chief of industrial operations."

Many CIOs worry that youth in America are not signing up for careers in IT. The answers to "What technology companies are you most familiar with?" shed some light on why: 34 percent said Microsoft, 31 percent said Apple, and 19 percent said Google, but only 3 percent said Dell and 1 percent said IBM.

When asked, "Do you consider technology a possible career path for yourself?," 28 percent said definitely or likely yes, 40 percent said definitely or likely no, and 25 percent said they were not sure.

The respondents were then asked, "As you seek your career, what is the most important thing to you?" Money was the top response, at 64 percent; 29 percent chose "making the most of your natural skills," and 28 percent chose making friends and being in a fun work environment.

Parents and immediate relatives were listed by 69 percent of survey participants as the group they would seek career advice from, with teachers and professors coming in second, at 40 percent.[75]

Michael Gabriel, the executive vice president and CIO of HBO, believes that the IT industry needs to do a much better job of marketing IT careers. According to Gabriel, "Most people are unfamiliar with the diversity of jobs that are available, as well as the lucrative compensation of those jobs. Most people don't know about the interesting and diverse skills needed to be a business analyst, a project leader, a business intelligence expert, or a portable device developer."

This misperception of IT careers begins at an early age, Gabriel said, adding, "Children don't know that they can actually create games for the iPad or Android device they are playing with. They don't see any IT role models, other than the classic geek in a TV show or movie. They see jobs such as doctors, criminologists, policemen, and firemen glorified on TV, but they have no idea what IT professionals do for a living."[76]

The survey results and Gabriel's comments reveal just how little the youth of America know about the IT industry. They do not know that jobs in IT, which require strong skills in STEM, are among the highest paying jobs in the United States.

But when you consider that 33 percent of the respondents had no idea what *IT* stands for and that 99 percent were not familiar with IBM, the results are not that surprising.

IT workers have a lot of work to do to address the misperceptions.

2009: *The Economic Impact of the Achievement Gap in America's Schools*

In this 2009 report, McKinsey & Company links an economic effect to U.S. student scores in international math and science assessment tests. It is the first ever such evaluation. McKinsey & Company claims it was necessary because "while great emphasis has been placed on educational achievement gaps, . . . its economic impact has received less attention."[77]

The Four Gaps

McKinsey lists four types of achievement gaps. A gap exists between the following:

1. The United States and other nations (mostly high-performing countries like Finland and [South] Korea).
2. Black and Latino students and white students in the United States.
3. Students of different income levels.
4. Similar students schooled in different systems and regions.[78]

A Larger U.S. GDP States If Only . . .

When the four gaps were evaluated together, McKinsey & Company concluded the following:

> This report finds that the underutilization of human potential in the United States is extremely costly. Our results show that short-falls in academic achievement—which are avoidable—impose heavy and often tragic consequences, via lower earnings, poorer health, and higher rates of incarceration; lagging achievement as early as fourth grade appears to be a powerful predictor of rates of high school and college graduation, as well as lifetime earnings;

if the United States had in recent years closed the gap in educational achievement levels with better performing nations such as Finland and [South] Korea, U.S. GDP in 2008 could have been $1.3 trillion to $2.3 trillion higher. This represents 9 to 16 percent increase of GDP.[79]

"A Permanent National Recession"

McKinsey & Company then dramatically lowers the boom on the educational gaps in America with this proclamation:

> Put differently, the persistence of these educational achievement gaps imposes on the United States the economic equivalent of a permanent national recession. The recurring annual economic cost of the international achievement gap is substantially larger than the deep recession the United States is currently experiencing.[80]

If One School Can Do It . . .

There was a silver lining in the McKinsey & Company report, however:

> While the price of the status quo in educational outcomes is remarkably high, the promise implicit in these findings is compelling. In particular, the wide variation in performance among schools and school systems serving similar students suggests that the opportunity and output gaps related to today's achievement gap can be substantially closed.[81]

The pragmatic consultant rationale was this: if one school can solve this thorny problem, so can another. It's a simple, hopeful, and powerful observation.

2009: *The Widget Effect: Our National Failure to Acknowledge and Act on Differences in Teacher Effectiveness*

> Who are the best teachers? The question is simple enough. There's just one problem: except for word of mouth from other parents, no one call tell you the answers.
>
> —DANIEL WEISBERG, *THE WIDGET EFFECT*[82]

You have now read about report after report touting the importance of teacher quality as a key to improving America's education system. In 2009 the New Teacher Project, a national nonprofit organization "dedicated to closing the achievement gap by ensuring poor and minority students get outstanding teachers," funded *The Widget Effect: Our National Failure to Acknowledge and Act on Differences in Teacher Effectiveness*, a study of 15,000 educators in Arkansas, Colorado, Illinois, and Ohio.

The report's title is derived from "the tendency of school districts to assume classroom effectiveness is the same from teacher to teacher."[83] Hence, teachers could be replaced like mechanical widgets, with no harm to the education process.

I'm Okay, and So Are You!

According to *The Widget Effect*, "at the heart of the matter are teacher evaluation systems, which in theory should serve as the primary mechanism for assessing such variations [among teachers], but in practice tell us little about how one teacher differs from any other."[84] The study reviewed two approaches to teacher evaluation: a binary approach (satisfactory or unsatisfactory) and a broader range of ratings. To underscore its point, *The Widget Effect* says that in school districts that use the binary evaluation rating, 99 percent of the teachers reviewed received a satisfactory evaluation. In the broader scale, 94 percent of the teachers reviewed received the top 2 (out of 10) ratings.

The provocative question raised by *The Widget Effect* is this: If all teachers are favorably reviewed, how can you identify bad teachers?

The Flawed Process of Teacher Evaluation

At your job, how are you evaluated? By whether you meet time and budget deadlines? By your performance? By customer satisfaction? Here's what *The Widget Effect* said about the teacher evaluation process:

> The flawed teacher evaluation process is exacerbated . . . by cursory evaluation practices and poor implementation. Evaluations are short and infrequent. . . . The result is clear: evaluation systems fail to differentiate among teachers. As a result, teacher effectiveness is largely ignored. Excellent teachers cannot be recognized or rewarded, chronically low-performing teachers languish, and the wide majority of teachers performing at moderate levels do not get the support and development they need to improve as professionals.[85]

Everyone Can't Be a Superstar

The Widget Effect continues to be a controversial report. I asked several of my friends who teach to describe the teacher evaluation process for me. Most described it exactly as *The Widget Effect* does. One friend, who entered teaching as a second career from the consulting business more than a decade ago, told me, "Most businesses have a clearly defined HR [human resource] system in place to quickly identify superstars. In education, we are all 'superstars,' told time and time again how good we are doing by our principals or department chairperson. Once you get inside the system, the motivation to improve, besides going to off-site development classes that are often perceived as vacation days, drops considerably."

2009: *Steady As She Goes? Three Generations of Students through the Science and Engineering Pipeline*

In 2000, the National Science and Technology Council, a group of experts appointed by the Clinton administration, had warned of a shortage of scientists and engineers in *Ensuring a Strong U.S. Scientific, Technical, and Engineering Workforce in the 21st Century*, affixing blame on an American education system that was inept at producing a diverse STEM workforce.

Nine years later, B. Lindsay Lowell of Georgetown University and Hal Salzman of the Urban Institute—whom we have already met as the authors of the 2007 report *Into the Eye of the Storm*—authored *Steady As She Goes? Three Generations of Students through the Science and Engineering Pipeline*. This study sought "evidence of a long-term decline in the proportion of American students with the relevant training and qualifications to pursue STEM jobs."[86]

What they found depends on which data set you examine—hence the question mark in the title of the report.

Three Key Career Gateways

Steady As She Goes? examined three key transition points in the STEM pipeline:

1. High school to a STEM degree in college (5 years after high school).
2. Completion of a STEM degree to first job (3 years after college).
3. Completion of a STEM degree to employment in a STEM occupation at midcareer (10 years after college).

Utilizing SAT scores, college transcripts, and employment data, *Steady As She Goes?* reports data from 1972 to 2005 in each of three STEM career transition periods (see Tables 11.15, 11.16, and 11.17).[87]

Table 11.15 Transition 1: High School Graduates Who Major in a STEM Discipline

Student Category	1977	1997	2005
All high school SAT takers	9.6%	10%	8.3%
Top quintile of SAT takers	21%	28.7%	13.8%
Bottom quintile of SAT takers	2%	1%	0%

Table 11.16 Transition 2: STEM College Graduates to First STEM Job (3 Years after College)

Student Category	1980	1996	2000
All STEM college graduates	31.5%	52.8%	44.9%
Top quintile college GPA students	38%	52%	41.8%
Bottom quintile college GPA	31%	51%	41%

Table 11.17 Transition 3: STEM College Graduates to STEM Job (10 Years after College)

Student Category	1987	1996	2003
All STEM college graduates	34.8%	38%	43.7%
Top quintile college GPA students	45%	44%	43%
Bottom quintile college GPA	23%	29%	42%

The STEM Turning Point

When examined long-term (30 years or more), the trends in high school, college, and STEM employment data are relatively unchanged. But if you narrow the evaluation period to the 1997–2005 time frame, a different picture emerges that suggests the brightest STEM workers are leaving STEM employment for other fields.

Lowell and Salzman explain it this way:

It appears that the 1990s marked a turning point in longer-term trends, at least for the best students in high school and college. The top quintile SAT/ACT and GPA performers appear to have been dropping out of the STEM pipeline at a substantial rate, and this decline seems to have come on quite suddenly in the mid-to-late

1990s. . . . This may indicate that the top high school graduates are no longer interested in STEM, but it might also indicate that a future in a STEM job is not attractive for some reason.

The decline in retention from college to first job might also be due to loss of interest in STEM careers, but alternatively top STEM majors may be responding to market forces and incentives. . . . Highly qualified students may be choosing a non-STEM job because these other occupations are higher paying, offer better career prospects, and are less susceptible to offshoring.[88]

The Rear Guard

Highly talented STEM college graduates were not the only category of STEM workers to show a dramatic change in trends from the mid-1990s to 2005. The bottom quintile of college STEM graduates made remarkable gains during the period (see Tables 11.18 and 11.19).[89]

Lowell and Salzman say the following about the upward composition trend of the bottom quintile among America's STEM workforce: "It does suggest we turn our attention to factors other than educational preparation or student ability in this compositional shift to lower-performing students in the STEM pipeline."[90]

In the first decade of the twenty-first century, the warnings about the inadequacies of the U.S. education system to teach science and math increased in number and seriousness as the U.S. skills gap widened. As the second decade dawned, in 2010, the warnings were no longer academic; they started to predict specific economic, workforce, and national security risks. The gap was about to become wider still.

Table 11.18   Bottom Quintile Employed in STEM Job 3 Years after College

Year	Percentage Employed
1980	31%
2000	41%

Table 11.19   Bottom Quintile Employed in STEM Job 10 Years after College

Year	Percentage Employed
1987	23%
2003	42%

Notes

1. C. S. Lewis, "The Magician's Nephew," in *The Chronicles of Narnia* (New York: HarperCollins, 1952).
2. National Science and Technology Council, *Ensuring a Strong U.S. Scientific, Technical, and Engineering Workforce in the 21st Century*, April 2000, www.whitehouse.gov/sites/default/files/microsites/ostp/workforcerpt.pdf.
3. Ibid.
4. U.S. Citizenship and Immigration Services, www.us-immigration.com.
5. Norm Matloff, Professor Norm Matloff's H-1B Web Page, http://heather.cs.ucdavis.edu/h1b.html.
6. Tommy Cornelis, "U.S. News Seeks Solutions to America's STEM Crisis," STEM Connector, May 18, 2012, www.stemconnector.org.
7. National Commission on Mathematics and Science Teaching for the 21st Century, *Before It's Too Late*, 2000, www.nationalmathandscience.org/sites/default/files/resource/before%20its%20too%20late.pdf.
8. Ibid.
9. Ibid.
10. Anne C. Kolarik, "The Percentage of Highly Qualified Math/Science Teachers by State before and after NCLB," PhD diss., University of Kansas, Lawrence, 2010.
11. National Commission on Mathematics and Science Teaching, *Before It's Too Late*.
12. Organization for Economic Co-Operation and Development, *Programme for International Student Assessment*, 2000, www.oecd.org/pisa/faqoecdpisa.htm.
13. National Center for Educational Statistics, *National Assessment of Educational Progress*, 2000, http://nces.ed.gov/annuals.
14. Ulysses Lee, *History of the Army* (U.S. Army: Washington, DC), www.history.army.mil/books/wwii/11-4.
15. David C. Berliner and Bruce J. Biddle, *The Manufactured Crisis: Myths, Fraud, and the Attack on America's Public Schools* (Reading, MA: Addison-Wesley, 1995), 33.
16. National Commission on Teaching and America's Future, *Unraveling the "Teacher Shortage" Problem: Teacher Retention Is the Key*, 2002, www.ncsu.edu/mentorjunction/text_files/teacher_retentionsymposium.pdf.
17. Thomas Carroll and Elizabeth Foster, *Who Will Teach?* (Washington, DC: National Commission on Teaching and America's Future, 2010), 10.
18. Ibid., 3.
19. Ibid., 13.
20. Ibid.
21. Business Higher Education Forum, *Building a Nation of Learners*, 2003, www.bhef.com/publications/documents/building_nation_03.pdf.
22. Ibid.
23. Ibid.
24. President's Council of Advisors on Science and Technology, *Sustaining the Nation's Innovation Ecosystem*, 2004, www.whitehouse.gov/sites/default/files/microsites/ostp/pcast-04-sciengcapabilities.pdf.
25. Ibid.
26. Niall Ferguson, "In China's Orbit: After 50 Years of Western Predominance the World Is Tilting to the East," *Wall Street Journal*, November 18, 2000.
27. National Science Foundation, "Science and Engineering Indicators," 2004, www.nsf.gov/statistics/seind04.
28. President's Council, *Sustaining the Nation's Innovation Ecosystem*.
29. Peter Drucker, "Drucker Sets Us Straight," *Fortune*, January 12, 2004.

30. American Electronics Association, *Losing the Competitive Advantage: The Challenge for Science and Technology in America* (Washington, DC: American Electronics Association, 2005), 5.
31. Ibid., 23.
32. Task Force on the Future of American Innovation, *The Knowledge Economy: Is the United States Losing Its Competitive Edge?*, February 16, 2005, www.nationalmathandscience.org/sites /default/files/resource/knowledge%20economy.pdf.
33. Ibid.
34. Thomas Friedman, *The World Is Flat: A Brief History of the 21st Century* (New York: Picador, 2005), 36.
35. Ibid., 48.
36. Norman R. Augustine, testimony to the Senate Committee on Energy and Natural Resources, October 2005.
37. National Academy of Sciences, *Rising above the Gathering Storm: Energizing and Employing America for a Brighter Economic Future* (Washington, DC: National Academies Press, 2005), www.nap.edu/openbook.php?record_id=11463&page=1.
38. Ibid.
39. Ibid.
40. Jeffrey L. Furman, *The America COMPETES Acts: The Future of U.S. Physical Science and Engineering Research?* (Washington, DC: National Bureau of Economic Research Innovation Policy & the Economy Workshop, 2012), 2.
41. National Center for Educational Statistics, Annual Reports, http://nces.ed.gov/annuals.
42. Educational Testing Service: *Teachers and the Uncertain American Future* (Princeton, NJ: Educational Testing Service, 2006), 1.
43. Ibid. 5.
44. Ibid., 18.
45. Willie Pearson Jr. and Alan Fechter (eds.), *Who Will Do Science?: Educating the Next Generation* (Baltimore: John Hopkins University Press, 1994).
46. Shirley Ann Jackson, *The Quiet Crisis: Falling Short in Producing American Scientific and Technical Talent* (Troy, NY: Rensselaer Polytechnic Institute, 2006), www.rpi.edu/homepage/ quietcrisis/index.html.
47. American Electronics Association, *We Are Still Losing Our Competitive Advantage: Now Is The Time to Act* (Washington, DC: American Electronics Association, 2007), 1.
48. Ibid.
49. Ibid., 6.
50. Andreas Schleicher, "The Lisbon Council," Lisbon, Portugal, 2005.
51. McKinsey & Company, "How the World's Best-Performing School Systems Come Out on Top," September 2007, www.mckinseyonsociety.com/how-the-worlds-most-improved- school-systems-keep-getting-better.
52. Ibid.
53. Ibid.
54. Ibid.
55. B. Lindsay Lowell and Harold Salzman, *Into the Eye of the Storm: Assessing the Evidence on Science and Engineering Education Quality and Workforce Demand*, Urban Institute, 2007, www.urban.org/publications/411562.html.
56. Ibid.
57. Ibid.
58. Ibid.
59. Ibid.
60. Commission on the Skills of the American Workforce, *Tough Choices or Tough Times: The Report of the New Commission on Skills of the American Workforce* (Rochester, NY: National Center on Education and the Economy, 2007), www.skillscommission.org/wp-content/ uploads/2010/05/ToughChoices_EXECSUM.pdf.

61. Ibid.
62. Ibid.
63. Ibid.
64. Eric Hanushek and Ludger Wolfmann, *The Role of Education Quality in Economic Growth*, World Bank Policy Research Working Paper 4122, February 2007, https://openknowledge .worldbank.org/bitstream/handle/10986/7154/wps4122.pdf?sequence=1.
65. Ibid.
66. Ibid.
67. Emmeline Zhao, "New York City Teacher Ratings: Teacher Data Reports Publicly Released Amid Controversy," *Huffington Post*, February 24, 2012, www.huffingtonpost .com/2012/02/24/new-york-city-rat_n_1299837.html.
68. National Mathematics Advisory Panel, *Foundations for Success: The Final Report of the National Mathematics Advisory Panel*, March 2008, www2.ed.gov/about/bdscomm/list/mathpanel/ report/final-report.pdf.
69. Ibid.
70. Louis V. Gerstner Jr., "Lessons from 40 Years of Education Reform," *Wall Street Journal*, December 1, 2008.
71. Rob Atkinson and Michael Shellenberger, *Rising Tigers, Sleeping Giant: Asian Nations Set to Dominate the Clean Energy Race by Out-Investing the United States* (Washington, DC: Breakthrough Institute and the Information Technology and Innovation Foundation, 2009), 3.
72. Ibid.
73. Ibid., 3.
74. Ibid., 10.
75. CIO Executive Council, *Youth and Technology Careers: Awareness, Attitudes and How to Influence* (Farmingham: MA: CIO Executive Council, 2009), 8.
76. Michael Gabriel, e-mail to author, August 17, 2012.
77. McKinsey & Company, *The Economic Impact of the Achievement Gap in America's Schools*, April 2009, http://mckinseyonsociety.com/the-economic-impact-of-the-achievement-gap-in- americas-schools.
78. Ibid.
79. Ibid.
80. Ibid.
81. Ibid.
82. Daniel Weisberg, *The Widget Effect: Our National Failure to Acknowledge and Act on Differences in Teacher Effectiveness* (New York: New Teacher Project, 2009), 3.
83. Ibid., 4.
84. Ibid.
85. Ibid.
86. B. Lindsay Lowell and Hal Salzman, *Steady As She Goes? Three Generations of Students through the Science and Engineering Pipeline* (Washington, DC: Georgetown University Institute for the Study of International Migration, 2009), 1.
87. Ibid., 30.
88. Ibid., 31.
89. Ibid., 17.
90. Ibid., 32.

CHAPTER 12

The Consequences of the Skills Gap Become Apparent

The future warns us through current symptoms.

—AUTHOR TOBA BETA[1]

2010: *Rising above the Gathering Storm Revisited: Rapidly Approaching Category 5*

In the face of so many other daunting near-term challenges, U.S. government and industry are letting the crucial strategic issues of U.S. competitiveness slip below the surface.

—NATIONAL ACADEMY OF SCIENCES,
RISING ABOVE THE GATHERING STORM REVISITED[2]

Rising above the Gathering Storm Revisited: Rapidly Approaching Category 5 was the second report issued by the National Academy of Sciences, a 2010 follow-up to the first report, *Rising above the Gathering Storm: Energizing and Employing America for a Brighter Economic Future*, released five years earlier. A comparison of the titles of the two reports tells a story: rather than "rising above" the storm in 2005, five years later America was on a path to fly directly into a category-five hurricane, the most dangerous hurricane ranking.

How Does This Scenario of America's Future Suit You?

Rising above the Gathering Storm Revisited offered a dire long-term view of America's problem, claiming, "It is widely agreed that addressing America's competitiveness challenge is an undertaking that will require many years, if not decades, to solve."

It described what had occurred since 2005 when it released its first warning, *Rising above the Gathering Storm*:

> Our nation's outlook has worsened, our overall public school system has shown little sign of improvement, particularly in mathematics and science, and many other nations have been markedly progressing, thereby affecting America's relative ability to compete effectively for new factories, research laboratories, administrative centers—and jobs.[3]

Batten Down the Hatches!

The National Academy of Sciences has earned "street cred" by issuing two reports on the state of American competitiveness within five years.

Therefore, the main conclusion of the new report deserves special attention:

> In balance it would appear that overall the United States long-term competitiveness outlook (read jobs) has further deteriorated since the publication of the "Gathering Storm" report five years ago. Today, for the first time in our history, America's younger generation is less well educated than its parents. For the first time in the nation's history, the health of the younger generation has the potential to be inferior to that of its parents. And only a minority of American adults believes the standard of living of their children will be higher than what they themselves have enjoyed.
>
> To reverse this foreboding outlook will require a sustained commitment by both individual citizens and by the nation's government . . . at all levels. The Gathering Storm is looking ominously like a Category 5 . . . and, as the nation has so vividly observed [from Hurricane Katrina], rebuilding from such an event is far more difficult than preparing in advance to withstand it.[4]

2010: *Why So Few Women in Science, Technology, Engineering, and Mathematics?*

Women are 51 percent of all Americans, and they hold 48 percent of the nation's jobs. But women account for only 24 percent of STEM positions in the United States, and only 17 percent of the IT executives who read *CIO* magazine are women.

These figures are certainly ominous news for women in the technology business. This trend pattern motivated the American Association of University Women (AAUW) to release its *Why So Few Women in Science, Technology, Engineering, and Mathematics?* report in 2010.

Women do hold prominent places in the history of computing. Ada Lovelace was recognized as the first computer programmer for Charles Babbage's analytical engine in the nineteenth century. Grace Hopper was the first programmer of the Mark I mainframe and is widely acknowledged as the "mother of COBOL." My favorite is Jean Bartik, who was the first programmer for the ENIAC computer and once commented that women working in the computer business should "look like a girl, act like a lady, think like a man, and work like a dog!"[5]

Despite these impressive lists of firsts for women in computing, and the incredible progress women have made in education and the workplace over the past 60 years, the STEM fields remain dominated by men.

The AAUW wanted to find out why. Here is a summary of its report.

Interest Lost in College

Although men outnumber women widely in technology professions, the AAUW found that in elementary school and high school, "girls and boys take math and science courses in roughly equal numbers, and about as many girls leave high school prepared to pursue science and engineering majors in college." The AAUW noted that "30 years ago there were 13 boys for every girl who scored above 700 in SAT math, and by 2010 that ratio has shrunk to about 3:1."

But then it seems that the STEM career switch is turned off in most college-age women. The AAUW report claimed that "by graduation men outnumber women in nearly every science and engineering field, and the difference is dramatic, with women earning only 20% of bachelor's degrees in STEM fields."[6]

The Early Years Count

The AAUW reported that the foundation for women entering STEM careers is laid early in life; it is mostly influenced by parents and teachers, who can instill a simple concept in the minds of young women: that they can be successful. According to the AAUW report, girls do better on math tests and are more likely to say they want to study math in the future if they receive this positive reinforcement from parents and teachers. Moreover, at a young age, encouraging young women in science and math works well to counter the powerful negative stereotypes about girls in STEM.

Must Do Better Than the Guys

The AAUW report offered another interesting insight: girls believe they aren't as smart as boys in subjects like math and science. Girls counterbalance that lower self-assessment by holding "themselves to a higher standard than boys in subjects like math, believing that they have to be exceptional to succeed in male fields. One result of girls' lower self-assessment of their math ability—even in the face of good grades and test scores—and their higher standards for performance is that fewer girls than boys aspire to STEM careers."[7]

Changes Colleges Can Make

"The foundation for a STEM career is laid early in life, but scientists and engineers are made in colleges and universities," the AAUW noted, so it made the following three recommendations to counterbalance the steep decline in the number of young women who earn STEM undergraduate degrees:

1. Professors must do a better job providing a broader view of STEM opportunities in introductory courses.
2. Colleges and universities must be more proactive in recruiting female faculty for STEM departments.
3. Because female STEM professors are often less satisfied with the rigid academic workplace, higher education institutions should be proactive and implement more mentoring and work-life programs for female STEM professors.

Nature or Nurture?

The AAUW report offered the following inclusive summation on the imbalance of the sexes in STEM:

> The striking disparity between the numbers of men and women in STEM has often been considered as evidence of biologically driven gender differences in abilities and interests. The classical formulation of this idea is that men naturally excel in mathematically demanding disciplines, whereas women naturally excel in fields using language skills. Recent gains in girls' mathematical achievement, however, demonstrate the importance of culture and learning environments in the cultivation of abilities and interests. To diversify the STEM fields we must take a hard look at the stereotypes and biases that still pervade our culture. Encouraging more girls and women to enter

vital fields will require careful attention to the environment in our classrooms and workplaces and throughout our culture.[8]

2010: *Waiting for Superman*

In 2010, a movie was released about the American education system. *Waiting for Superman*, directed by Davis Guggenheim (who also directed Al Gore's documentary, *An Inconvenient Truth*), was a celebration of the charter school movement.

Critic David Kaplan said in his review of the film, "It is a well-intentioned cry for reform in American public education."[9] The film follows five students in their efforts, and those of their parents, to secure admission to a charter school, whose only entrance requirement is the luck of the draw—that is, winning a random lottery. This process creates incredible pressure for those who are waiting to learn of their fate. *Waiting for Superman* does an excellent job of conveying that pressure and is worth seeing, particularly if you are curious about the charter school movement.

Other excellent films on the state of American education are *October Sky* (1990), *Two Million Minutes* (2008), *Race to Nowhere* (2010), *The Finland Phenomenon* (2011), and *Won't Back Down* (2012).

2010: *Education Next's* Public Perception of Education Survey

Education Next is an academic journal with the following mission:

> In the stormy seas of school reform, this journal will steer a steady course, presenting the facts as best they can be determined, giving voice (without fear or favor) to worthy research, sound ideas, and responsible arguments.[10]

So, "without fear or favor," in 2010, *Education Next* decided to poll a national sample of parents and teachers on how the public perceived the state of public education in the nation. Here are the major findings, presented as questions and tables that supply the answers.

Question to parents: Students are often given the grades A, B, C, D, and F to denote the quality of their work. Suppose the public schools themselves were graded in the same way. What grade would you give the public schools in the nation as a whole? (See Table 12.1.)

Table 12.1   Parental Perception of U.S. Public Education System

Grade	Percent
A	1%
B	17%
C	55%
D	21%
F	6%

Source: *Education Next*, Program on Education Policy and Governance Survey 2010, http://educationnext
.org/files/complete_survey_results_2010.pdf.

Parents were then presented with the names of their local elementary school and middle school from a list of schools matched to their home zip codes and asked the same question.

Question to parents: How would you grade your elementary and middle schools? (See Table 12.2.)

Question to parents and teachers: Do you think that local taxes to fund public schools should increase, decrease, or stay about the same? (See Table 12.3.)

Table 12.2   How Parents Rate Their Local Schools

Grade	Elementary School	Middle School
A	18%	16%
B	47%	39%
C	28%	33%
D	5%	9%
F	2%	3%

Source: *Education Next*, Program on Education Policy and Governance Survey 2010, http://educationnext
.org/files/complete_survey_results__2010.pdf.

Table 12.3   What Should Happen to Local Taxes to Fund Schools

Taxes Should . . .	Parents	Teachers
Greatly increase	5%	5%
Increase	27%	40%
Stay the same	55%	47%
Decrease	11%	5%
Greatly decrease	2%	2%

Source: *Education Next*, Program on Education Policy and Governance Survey 2010, http://educationnext
.org/files/complete_survey_results__2010.pdf.

Table 12.4   Would More Money Result in Better Education?

Confidence Level	Parents	Teachers
Very confident	20%	19%
Somewhat confident	41%	43%
Not very confident	27%	28%
Not confident at all	12%	11%

Source: *Education Next*, Program on Education Policy and Governance Survey 2010, http://educationnext
.org/files/complete_survey_results__2010.pdf.

Question to parents and teachers: If more money were spent on public schools in your district, how confident are you that students would learn more? (See Table 12.4.)

Question to parents: Some people have proposed that states be required to toughen the standardized tests used to evaluate student performance. Do you support or oppose this proposal? (See Table 12.5.)

Here's my attempt to distill all this interesting data: (1) 82 percent of American parents give the public schools (overall) a grade of C or lower, (2) they believe that the quality of their local middle school (to which 45 percent give a C or lower) is less than that of their local elementary school (to which 35 percent give a C or lower), (3) 45 percent of the teachers in the sample claim that local taxes to support public schools should increase/greatly increase, yet 39 percent of those teachers have little or no confidence that more money would lead to better education, and (4) only 33 percent of teachers support standardized testing.

The pain of a weak public school system has not reached a crisis point. Teachers want more taxes but cannot promise a better education as a result.

Table 12.5   How Parents Feel about Standardized Testing

Support Level	Parents	Teachers
Completely support	19%	9%
Somewhat support	35%	24%
No opinion	22%	22%
Somewhat oppose	17%	32%
Completely oppose	6%	14%

Source: Program on Education Policy and Governance Survey 2010, http://educationnext.org/files/
complete_survey_results__2010.pdf.

I remain concerned about the slippage from elementary to middle school. According to the Department of Education, the slippage is even worse for high school.

2010: Interview with Craig Barrett

"This is a disaster," said Craig Barrett, the former chairman of the Intel Corporation and a key member of the National Academy of Sciences committees that issued the two "storm" reports mentioned earlier. After his testimony on Capitol Hill about the *Rising above the Gathering Storm Revisited* report, Barrett was interviewed on Bloomberg Television. This is what he said:

> There is no silver bullet to fix the education program. The real challenge is we've been warning about this problem since the early 1980s. We've been 30 years into this exercise of saying our K through 12 system is broken. We have too low expectations. We're not getting kids interested in math and science. There's a whole litany of problems and we just haven't done anything about them. We already spend more than anybody in the world on education. If you look at every positive education system around the world, every successful one has three basic features. It has great teachers. It has high expectations of the kids. And it has a good feedback loop to help teachers or kids who are struggling. If you look at the United States, we draw our teachers usually from the bottom 25 percent of our college graduates. Most folks who teach science in our schools are not graduates of science, not accredited. Only two-thirds of the people who teach math are accredited. We have the lowest expectations in the world of our students. Every time we put a feedback loop in place, like No Child Left Behind, it gets politicized. The teacher unions complain about being measured, the insult of being measured. Administrators complain about it, the insult of being measured. We're not facing reality in any of these topics. We're really zero for three in the characteristics you need for a solid education system. We need public support. We need public recognition. The real challenge is the lack of a coordinated national response. In 1983 "A Nation at Risk" report came out saying, "This is a disaster." The nation's governors in the 1990s said, "This is a disaster," the Glenn Commission in the 1990s said, "This is a disaster," and the National Academy of Sciences in 2005 said, "This is a disaster." You've got all the warning signals. We just have not responded.[11]

2010: *Closing the Talent Gap: Attracting and Retaining Top-Third Graduates to Careers in Teaching*

> People know that if you've been trained as a teacher you must really be something special.
>
> —FINLAND EDUCATION EXPERT[12]

In 2007, McKinsey & Company released the report *How the World's Best-Performing School Systems Come Out on Top*. It claimed that of "all the controllable factors in an education system, the most important, by far, is the effectiveness of the classroom teacher."[13]

The Formula for Success: Hire the Smartest

In 2010, McKinsey & Company issued a follow-up report, *Closing the Talent Gap: Attracting and Retaining Top-Third Graduates to Careers in Teaching*, that delved deeper into the main finding of the 2007 report: that recruiting teachers from the best of the best student population is the key to success. The new report claimed the following:

> The U.S. does not take a strategic or systematic approach to nurturing teaching talent. Buffeted by a chaotic mix of labor market trends, university economics, and local school district and budget dynamics, we have failed to attract, develop, reward, or retain outstanding professional teaching talent on a consistent basis. Fortunately, improving teacher effectiveness to lift student achievement has become a major reform theme in American education.[14]

Where Do the Best-Performing Countries Find Teachers?

McKinsey & Company studied the education systems of Finland, South Korea, and Singapore, three countries that continually perform well in international math and science assessment tests. Here's where those countries recruit their teachers:

- In Finland, teachers are recruited from the top 20 percent of high school graduates who go to college. One in 10 applicants for a teaching job is accepted.
- In South Korea, teachers are recruited from the top 5 percent of high school graduates who go to college.

- In Singapore, teachers are recruited from the top 30 percent of high school graduates who go to college. One in 8 applicants to a teaching position is accepted.[15]

The U.S. Mantra: "Give Me Money, That's What I Want!"

The United States has no national strategy to proactively recruit teachers from the top tier of college graduates. McKinsey claimed that in the United States today, "only 9% of top-third college students say they plan to go into teaching." McKinsey queried the 91 percent who weren't planning to become teachers to determine why they weren't considering teaching as a profession.

The top three responses were as follows: 90 percent said that teaching jobs do not offer a competitive starting salary, 87 percent claimed that if they chose teaching and did well, they wouldn't be rewarded financially, and 83 percent said that the skills they possessed would not be compensated appropriately.[16]

They'd Rather Be Garbage Collectors

Compared to South Korea, where the proverb "Don't step on even the shadow of a teacher" shows how much teachers are respected, in the United States the perception of the teaching profession held by college students is much lower. *Closing the Talent Gap* stated that "more than half [of college students in the survey] believe they could earn more as a garbage collector."[17]

Experience Counts

A Singapore education official told McKinsey & Company, "We believe the experience of a teacher is a very valuable asset to retain in the profession." In the United States, 46 percent of new teachers leave the profession within five years. Table 12.6 shows the annual teacher turnover rates for the United States, Finland, South Korea, and Singapore.[18]

Table 12.6   Teacher Turnover Rates

Country	Annual Rate
United States (high-poverty schools)	20%
United States (all schools)	14%
Singapore	3%
Finland	2%
South Korea	1%

So, America, What's the Game Plan?

McKinsey & Company stated clearly that recruiting and retaining the smartest people to teach is the key to success. In America, with a paltry 9 percent of the top-third college students opting for a career in teaching, and a national education system suffering a turnover rate that is almost 2,000 percent higher than that of other countries, what does our country need to do?

Closing the Talent Gap suggested the following six solutions—which would take decades to implement—to America's problem:

1. Make admissions to teacher training highly selective.
2. Pay for teacher education, as Singapore and Finland do (in the United States, students aspiring to be teachers often go into debt).
3. Apply the most rigorous teacher selection standards to those seeking to become elementary schoolteachers (in the United States, elementary school teachers are the least likely to come from the top third).
4. Improve the professional working environment for teachers (in the United States, professional development is often unprofessional, and opportunities for advancement and recognition are few).
5. Pay teachers a competitive salary (in the United States, top American college students see teacher pay as unattractive).
6. Give teachers more respect; top-performing nations accord enormous cultural respect to teaching and teachers.[19]

Who Should Teach?

The nineteenth-century American writer and satirist H. L. Mencken famously said, "Those who can, do. Those who can't, teach."

McKinsey & Company concluded its report with a much more provocative comment: "With much of America's teacher corps turning over in the next decade, the nation should be asking, 'Who should teach?'"[20]

The overwhelming recommendation: America's best and brightest.

2011: The National Assessment of Educational Progress

The 2011 NAEP test measured mathematics and science scores for America's eighth-grade students, and across the board there were positive results. Eighth-grade scores in math (see Table 12.7) and science (see Table 12.8) improved markedly from 2005. Could it be that mandatory testing was starting to have a positive effect on student performance?[21]

Table 12.7   Eighth-Grade Mathematics NAEP Scores

Year	Advanced	Proficient	Basic	Below Basic
2005	5%	30%	44%	21%
2011	11%	33%	39%	16%

Table 12.8   Eighth-Grade Science Scores

Year	Advanced	Proficient	Basic	Below Basic
2005	3%	24%	30%	43%
2011	2%	29%	34%	36%

Even though 31 percent of eighth-grade students were proficient or better in science and 44 percent were proficient or better in math, both percentages are far short of the mandated goal of the No Child Left Behind Act that 100 percent of U.S. students be proficient (or better) in science and math by 2014.

By early 2012, 36 states knew they couldn't possibly meet that mandate and filed for exemptions from it.

2011: The Intel Corporation's Survey of Teens' Perceptions of Engineering

You know the saying: Perception is reality. In 2011, the Intel Corporation was curious about how teenagers (ages 13–18) perceive careers in engineering. The study, conducted across a sample of 1,004 teens with computer access, reported these findings:

1. Teens view engineers as "inventive" and "smart," but the word they associated most with engineering as a career was *difficult*.
2. 63 percent of the teens said they "have not considered a career in engineering." Of those considering the profession, 74 percent are doing so because they think it "would be interesting." Moreover, twice as many boys (37 percent) as girls (18 percent) say they have considered a career in engineering.
3. The teens said that a career in engineering pays well, offers job security, and often provides an opportunity to change the world, but familiarity with what an engineer does was weak compared to the professions the teens said they were most familiar with: teacher, doctor, nurse, police officer, and chef.

4. 44 percent of the teens said that more exposure to engineering would make them reconsider engineering as a career.[22]

The Intel study findings were interesting but not unexpected. Moreover, the fourth finding underscores the responsibility that every IT worker has in spreading the word about what a great career one can have in a STEM field.

2011: *Globally Challenged: Are U.S. Students Ready to Compete?*

America faces many challenges . . . but the enemy I fear most is complacency. We are about to be hit by the full force of global competition. We must now establish a sense of urgency.

—CHARLES VEST, FORMER PRESIDENT, MASSACHUSETTS
INSTITUTE OF TECHNOLOGY[23]

In 2011, Harvard University's Program on Education Policy and Governance released *Globally Challenged: Are U.S. Students Ready to Compete?* This report was based on an analysis of the mathematics results of the 2009 Programme for International Student Assessment (PISA) exam. The report, similar to the 2009 McKinsey & Company report *The Economic Impact of the Achievement Gap in America's Schools*, addresses the economic consequences of the United States performing so poorly in the PISA exams.

What America Needs to Do

Globally Challenged claimed the following:

The United States could enjoy a remarkable increment in its annual GDP growth per capita by enhancing the math proficiency of U.S. students. Increasing the percentage of proficient students to the levels attained in South Korea [South Korea came in fourth in the 2009 PISA study with a math proficiency score of 546 while the United States claimed thirty-first place with a score of 487] would increase the annual U.S. growth rate by 1.3 percentage points. Since long-term average annual growth rates hover between two and three percentage points, that increment would lift growth rates by 50%.

When translated into dollar terms, these magnitudes become staggering. If one calculates these percentage increases as national income projections over an 80-year period (providing for a 20-year delay before any school reform is completed and newly proficient students begin their working careers), a back-of-the-envelope calculation suggests gains of nothing less than $75 trillion over the period. Those who say student math performance does not matter are clearly wrong.[24]

Can American Students Compete with Rest of the World?

Globally Challenged asked a critical question: Are U.S. students ready to compete? Vint Cerf, one of the key inventors of the Internet, said in the report, "America simply is not producing enough of our own innovators, and the cause is twofold—a deteriorating K–12 education system and a national culture that does not emphasize the importance of education and the value of engineering and science."[25]

Skills Gap on the Horizon: Two Million Workers Needed

Globally Challenged cited a McKinsey Global Institute study that "estimates over the next few years there will be a gap of nearly two million workers with the necessary analytical and technical skills" for a twenty-first-century global workforce that puts a premium on service skills.[26]

The most interesting fact of this Harvard report to me was the projection that if political and education leaders in the United States woke up tomorrow and committed themselves to systemic national education reform, it would still take 20 years before our country realized the results of that decision.

Do Americans have that kind of patience? I think not.

2012: How Well Are American Students Learning?

The Brown Center on Education Policy at the Brookings Institute released *How Well Are American Students Learning?* in February 2012. The report reviewed three issues: (1) the Common Core State Standards, (2) achievement gaps in the NAEP tests, and (3) the misinterpretation of international test scores referred to earlier in this book.

The Common What?

The Common Core State Standards Initiative is a project sponsored by the Council of State School Officers and the National Governors Association. The Common Core aims to determine what students should learn in mathematics, science, and other subjects in grades K–12. According to the report, "the standards are written by teams of curriculum specialists and vetted by academics, teachers, and other experts."[27]

The initiative was funded by the Department of Education in 2010 with the goal of delivering a set of education standards that could be tested by 2014. Yet when the Department of Education was founded in November 1979, the document that created the department explicitly stated that the federal government was *not* to be a partner in setting a national curriculum agenda.

Why the Change Now?

Bill Gates is a supporter of the Common Core Initiative. Commenting on the need for a national curriculum agenda in the *Wall Street Journal*, he said, "It's ludicrous to think that multiplication in Alabama and multiplication in New York are really different."[28]

The Fordham Institute agreed with Gates, and in a report entitled *The Proficiency Illusion*, it questioned why America needs "fifty different assessments" and "fifty different definitions of what constitutes acceptable performance. How can a school from one state be labeled a failure while a school in another state, and with almost exactly the same test scores, can be considered a success?"[29]

Those supporting the Common Core Initiative claim it will set higher educational standards than current state standards. Others say a common national core curriculum will save hundreds of millions of dollars by consolidating the "many different versions of textbooks that are published to conform to individual states' curricular tastes."[30]

Opponents of Common Core make one strong argument. Their reasoning is that "the proposed common standards would undermine the decentralized, federal principles on which education has been governed since America's founding." Furthermore, "a one-size-fits-all centrally controlled curriculum makes no sense and only weak evidence supports the push for national standards."[31]

Common Core: An Uncommonly Bad Idea

How Well Are American Students Learning? concluded as follows:

> The analysis suggests [that a Common Core national curriculum] will have very little impact. The debate is sure to grow in intensity. It is about big ideas—curriculum and federalism. The proper role of the federal government and states, local districts, etc., especially in deciding what should be taught, remains a longstanding point of dispute. Standards with real consequences, like the No Child Left Behind Act, are most popular when they are first proposed. Their

popularity steadily declines from there. The Common Core will have little effect on American students' achievement. The nation will have to look elsewhere for ways to improve its schools.[32]

2012: *U.S. Education Reform and National Security*

By almost every measure, U.S. schools are failing to provide the kind of education our society will need to ensure American leadership in the 21st century. In short, America's failure to educate is affecting its national security.

—COUNCIL ON FOREIGN RELATIONS, U.S. EDUCATION
REFORM AND NATIONAL SECURITY [33]

The 2012 report from the Council on Foreign Relations, *U.S. Education Reform and National Security*, reviewed the linkage between a strong education system and U.S. national security, an issue first addressed 13 years earlier by the report *New World Coming*. Joel I. Klein, the former chancellor of the New York City Department of Education, and Condoleezza Rice, a secretary of state in the George W. Bush administration, chaired this effort.

Global National Security Concerns

Noting that "the domestic consequences of a weak education system are relatively well-known, Council on Foreign Relations focused its efforts on another issue: the national security of the United States and how it was being compromised by the weak education system. The report warned that "the United States must produce enough citizens with critical skills to fill the ranks of the Foreign Service, the intelligence community, and the armed forces."[34]

A "Menacing Threat to U.S. National Security"

Here's something fascinating. Remember our discussion in Chapter 1 of the Army General Classification Test, which gauged the mental capacity of men who wanted to join the armed forces in 1941? In 2012, 71 years later, the Council on Foreign Affairs report lamented the following:

Many U.S. generals caution that too many new enlistees cannot read training manuals for technologically sophisticated equipment. A former head of the Army's Training and Doctrine Command said that the lack of fully qualified young people was "an imminent and menacing threat to our national security."[35]

America: The Choice Is Yours

U.S. Education Reform and National Security concluded with this blunt challenge to America: "The Task Force believes that this country has a real, but time-limited, opportunity to make changes that would maintain the United States' position in the world and its security at home."[36]

2012: *Prosperity at Risk: Findings of Harvard Business School's Survey on U.S. Competitiveness*

71% of respondents expect U.S. competitiveness to decline over the next three years, with workers' living standards under greater pressure than firms' success.

—MICHAEL E. PORTER AND JAN W. RIVKIN, *PROSPERITY AT RISK*[37]

In June 2009, as the U.S. economy was recovering from the Great Recession of 2007, the Harvard Business School had an idea: to assess what impact the recession was going to have on the long-term competitiveness of the U.S. economy.

A Little Help from My Friends

The Harvard Business School had the right respondents for its survey: a network of 78,000 alumni who live and work in all regions of the world. The school sent a survey to all of them. Nearly 10,000 responded, including 1,700 respondents who were "personally involved in decisions about whether to place business activities and jobs in the U.S. or elsewhere."

The survey results were sobering: 71 percent of the respondents expect the U.S. competitiveness to deteriorate, "with firms less able to compete, less able to pay well, or both. Another 14% were neutral, and 16% were optimistic, expecting U.S. competitiveness to improve."[38]

Why the United States Is Uncompetitive

The report reviewed six reasons that America's competitiveness model is broken:

1. The U.S. macroeconomic policy is broken.
2. Political gridlock in Washington, D.C., prevents the passage of effective laws.
3. The United States has ineffective intellectual property rights laws, which make it easier for others to steal our ideas.

4. The U.S. legal framework is inefficient and very costly.
5. The national tax code is too complex.
6. The U.S. education system does not do an adequate job preparing students for productive work.[39]

Moving Time

In the survey, 43 percent of the respondents claimed that their firms made at least one decision about relocating a portion of the business or starting new activities in the past year, and nearly 60 percent of these relocation or start-up decisions included the possibility of moving the activity outside the United States. Only 9 percent said that those decisions involved moving an existing offshore component of the business back to the United States.[40]

Skills to Be Moved Offshore

Let's focus on job skills related to math and science. The Harvard Business School study did just that, with "research, development, and engineering skills." Even then, 42 percent of respondents reported that their firms had discussed moving jobs out of the United States, and firms planning to relocate 1,000 or more jobs were at the top of the list.[41]

Now and Then

So where is the United States headed? Only 57 percent of the respondents in the Harvard study said that the United States was doing "somewhat better" than other countries now. But the results shifted when U.S. prospects were compared in the future to advanced economies and emerging economies— then 70 percent said the United States would "keep pace" with "other advanced economies," and 66 percent said America would fall behind the economies of "emerging countries."[42]

Holding Off on Commitment

The results of the Harvard Business School study are interesting but not earthshaking. However, when you factor in the power-broking positions of the men and women who responded to the survey—a who's-who list of global business leaders, the people who make the big decisions on budgets, business locations, and human resource planning—their very tentative commitment to keep their businesses in America is of concern. Unless the United States fixes the six problem areas mentioned earlier, we can expect more relocation of American businesses overseas in the coming decade.

2012: The World Economic Forum's Annual *Global Competitiveness Report*

In September 2012, the World Economic Forum said the following in a press release it issued summarizing the findings of its annual report on U.S. competitiveness: "Despite increasing its overall competitiveness score, the United States continues its decline for the fourth year in a row, falling two more places to seventh position."[43]

The World Economic Forum is a well-respected international nonprofit agency that engages world leaders to discuss and shape key international economic, technological, academic, and health issues. The press release for *Global Competitiveness Report, 2012–2013* included a list of the top 10 countries with a comparison of their scores from 2011 to 2012. Next to each country the report writers assigned a green, yellow, or red arrow to designate in which direction that country was heading.

The United States Heads in the Wrong Direction

Table 12.9 shows the comparison of the top 10 countries from 2011 to 2012, listed in order of their positions in the current report.[44]

The World Economic Forum assigned a red arrow pointing downward to the United States and offered this explanation: "In addition to the burgeoning macroeconomic vulnerabilities, some aspects of the country's institutional environment continue to raise concern among business leaders, particularly the low public trust in politicians and a perceived lack of governmental efficiency."[45]

Table 12.9   The United States Continues to Slide in Global Competitiveness

Country	2011 Ranking	2012 Ranking
Switzerland	1	1
Singapore	2	2
Finland	4	3
Sweden	3	4
Netherlands	7	5
Germany	6	6
United States	5	7
Britain	10	8
Hong Kong	11	9
Japan	9	10

Some critics are fond of noting that international comparisons, such as in this report, often put smaller, much less economically powerful countries ahead of the United States, and these critics repeat their favorite retort, "Who cares?" Indeed, the countries at the top of the 2012–2013 World Economic Forum report are indeed small countries that do not compete broadly with the United States.

But Germany does compete with the United States. And even though Germany remained in sixth place in this study, it moved ahead of the United States in 2012. Also moving up the list that year was Britain and Hong Kong, two entities the United States does compete with in many markets. Possibly the only economic comfort the United States can take out of this report is that Japan dropped one slot. All in all, it's not a good competitiveness report card for America.

Not a Great Comparison

The *Global Competitiveness Report* is based on cumulative rankings of 110 items. Table 12.10 shows a comparison of the 2011–2012 and 2012–2013 reports in 13 of those items.

Table 12.10   U.S. Cumulative Rankings for Two Years in 13 Items

Item	2011 Position (2011–2012 Report)	2012 Position (2012–2013 Report)
Market size	1	1
Availability of scientists and engineers	4	5
Labor market efficiency	4	6
Capacity for innovation	7	7
Business sophistication	10	10
Availability of latest technologies	18	14
Technological readiness	20	10
Financial market development	22	16
Infrastructure	24	14
Quality of education	26	28
Quality of science and math teachers	51	47
Soundness of banks	90	80
Government budget percentage of GDP	139	140

As I was developing that list, I thought of an imaginary press release on the overall condition of the United States based on those items. It could read like this:

> The United States slipped to seventh place in the World Economic Forum's *Global Competitiveness Report* released today. Though having the world's largest domestic market, the country fell for several key reasons: (1) despite the access to the availability of the latest technologies, the country failed to expand its capacity for innovation; (2) although the stability of its financial markets stabilized in the past 12 months, the government's budget as a percent of overall GDP remains a key concern; (3) business leaders have not shown a willingness to increase the sophistication of their business models, particularly in terms of labor market efficiency; (4) the quality of the country's education system, despite an annual investment of $583 billion each year, fell two places, to number 28; and (5) the country's availability of scientists and engineers fell again, leaving some to speculate that maybe the country's forty-seventh place in quality of math and science teachers was starting to create problems. When asked by this reporter for a comment, political leaders pointed to the failed policies of their adversaries as the primary reason for the country's slippage.

Drowning in a Sea of Mediocrity

In 1983, *A Nation at Risk* reported that America was engulfed in a "rising tide of mediocrity." That tide has risen for 30 years, and our country is now drowning in it.

2012: *Where Will All the STEM Talent Come From?*

"The supply of STEM talent is growing rapidly in some places," stated an Accenture Institute for High Performance report on the global STEM talent pool. "The key is learning how to find and access critical skills in the emerging global labor market."

This report, *Where Will All the STEM Talent Come From?*, makes ominous predictions about the American STEM workforce of the future. "Scientists, technologists, engineers, and mathematicians," the report claimed, "are the high-end knowledge workers of the global economy, and many business leaders fear shortages of STEM talent in the coming years."

The main point of the report was this:

> Our research suggests that many STEM talent shortages are lo-
> cal—not global. A global abundance of talent amid local short-
> ages means that employers will have to contend with location
> mismatch: STEM talent is available but not always in the places
> where it is needed. What we are witnessing is the emergence of
> a truly global labor market for STEM talent—but one that lacks
> essential mechanisms for matching demand and supply of critical
> skills across geographic boundaries. Many CEOs see the challenge
> of locating and forecasting talent availability as a major hurdle to
> growth.[46]

Sputnik, Part Two

The Accenture researchers then built a case for the concept of "location mis-
match" and what developed economies need to do:

> It is highly unlikely that universities in developed countries will be
> able to simply turn up the dial and produce more STEM graduates
> in the next decade. Consider that the United States graduated 88,000
> visual and performance arts majors in 2008 but only 69,000 engi-
> neers. The number of STEM graduates in the U.S. would need to
> increase by 20 to 30 percent by 2016 to meet the country's projected
> growth in science and engineering employment alone. That would
> require a collective effort on par with the one fueled by the space race
> between the United States and the Soviet Union in the late 1950s
> and 1960s.[47]

Yuasa's Phenomenon, Part Two

Yuasa's Phenomenon said that one of the key components of a world scientific
center is a large concentration of scientists and engineers. Table 12.11 shows
the trend numbers from Accenture between developed economies (e.g., United

Table 12.11   Percent of STEM Talent by Region of World

Year	Developed Economies	Emerging Economies
2013	13%	87%
2014	13%	87%
2015	12%	88%

Table 12.12   Percentage of STEM Degrees

Country	Percentage of Degrees
China	41%
India	26%
Britain	22%
Japan	18%
Brazil	14%
United States	13%

States, Britain, Japan) and emerging economies (e.g., Brazil, China, India). The data in the Accenture report reflect percentages of STEM degrees in individual countries, not the total globally.[48]

Accenture then analyzed the proportion of all degrees conferred in 2011 in a specific country that were STEM degrees (see Table 12.12).[49]

China outproduces the United States in STEM degrees by more than three to one. India beats the United States by two to one, and even Britain outpaces the United States in this measurement by 69 percent. Looked at another way, China, Brazil, and India collectively produce 81 percent of the world's new STEM workforce each year.

Where Will All the STEM Talent Come From? underscores the significant STEM numbers disadvantage facing the U.S. workforce in the future. The United States will never produce more STEM workers than China, India, or Brazil.

For the United States to remain a viable, economic superpower, it must make sure that the 1 in 10 STEM global workers produced by our education system annually is among the world's best. This is a foreboding challenge if America's underperforming trends in international math and science examinations are factored into the equation.

The data reflected in Accenture's report indicate that Yuasa's Phenomenon may be in play once again.

2012: SAT and ACT Scores Reveal Disappointing News

The year 2012 was not a good one for the 1,660,000 American students who took the SAT. Critical reading dropped to an average score of 496, the lowest since 1972. Writing dropped to 488, the lowest on record.[50] Math was the lone subject area that did not decline.

But the math scores didn't improve, either. In fact, they have shown little improvement in the past 18 years (see Table 12.13).[51]

Table 12.13   Minuscule Gain in SAT Math Score

Year	SAT Math Score
1996	511
2012	514

Ready or Not, Here I Come!

To recap: The SAT is now composed of three sections—math, critical reading, and writing. Each section is scored from 200 to 800, with 2,400 the highest possible score. In 2012, the College Entrance Examination Board introduced a new measurement designed to help states and private institutions to determine, based on an individual's SAT scores, the likelihood of college success. The new tool is described in an annual report called *The SAT Report on College and Career Readiness*.

Here's how it works. The College Entrance Examination Board simply adds up the scores from the three sections of the SAT. The SAT College and Career Readiness benchmark is 1,550 points. If a student's total score is 1,550 or higher, the College Entrance Examination Board predicts a 65 percent probability that the student will earn a B average or higher in core college courses in freshman year.

Sounds reasonable, right? And it is.

The problem is that in 2012, using this measurement, the College Entrance Examination Board predicted that only 43 percent of high school seniors who took the SAT were ready to do college work, the same percentage reported in 2011. Gaston Caperton, president of the College Entrance Examination Board, issued a press release announcing the dismal 2012 college readiness results:

> This report should serve as a call to action to expand access to rigor for more students. Our nation's future depends on the strength of our education system. When less than half of kids who want to go to college are prepared to do so, that system is failing. We must make education a national priority and deliver more rigor to students.[52]

A Hard ACT to Follow

As mentioned earlier, the SAT, launched in the 1920s, got a new competitor in the 1950s called the ACT exam. And even though the College Entrance Examination Board launched its college and career readiness benchmark in 2011, ACT was actually the first testing organization to launch such a measurement report entitled *Condition of College and Career Readiness* in 2008 (prior year reports can be located at www.act.org/readiness).

Table 12.14 College Readiness of U.S. Students

Subject	College Readiness Benchmark
English	67%
Reading	52%
Mathematics	46%
Science	31%
All four subjects	25%

The 2012 ACT's *Condition of College and Career Readiness* report uses four benchmarks: English composition, reading (social sciences), mathematics (college algebra), and science (biology). The benchmarks are ACT's proprietary measurements, extrapolated from ACT's score databases, that project to a student having a 50 percent chance of obtaining a B grade or higher or a 75 percent chance of earning a C grade or higher in a first-year college course. Table 12.14 shows how ACT ranked the college readiness of the 1.2 million test takers.[53]

Where's the Return on Investment?

The College Entrance Examination Board claims that 57 percent of students graduating from high school are not ready to succeed in college. ACT claims the number is higher: 75 percent. Those numbers are incredible! After 13 years of schooling in American classrooms—at a total cost of just over $136,000 per student—that many students have not learned the skills to succeed at college work is mind-blowing.

Let's put those numbers in a different context and equate them with customer satisfaction, a benchmark many businesses use to gauge current and future business success. Could your firm succeed if 66 percent (the average of the SAT and ACT percentages) of your customers were not satisfied with the product or service your company offered? How long would your corporation continue to exist with a number like that?

Now apply that to the United States and ask the same questions. A sobering thought, isn't it?

2012: *Five Misconceptions about Teaching Math and Science: American Education Has Not Declined, and Other Surprising Truths*

In June 2012, *Slate* magazine, a provocative online publication, published an entire issue on STEM education that featured an article written by David E. Drew, a professor and a former dean at Claremont Graduate University in Claremont, California, and the author of a 2011 book, *Stem the Tide: Reforming Science, Technology, Engineering, and Math Education in America*. In *Slate*, Drew addressed five misconceptions commonly held about STEM education. (These misconceptions have been addressed in this book and will be discussed further in the Epilogue.)

Misconception 1: American Schools Have Only Recently Declined

Drew wrote the following:

> Fact: the mantra from many educators and policy-makers for a quarter-century has been to lament the decline of American schools. Even the classic 1983 report "A Nation at Risk," which sounded the alarm about the American education system says, "What was unimaginable a generation ago has begun to occur—others are matching and surpassing our educational attainments." But this is a flawed assessment of our past.
>
> The fact that we score poorly now does not mean that our educational system has deteriorated. In fact, it was always bad. Our high school students have always scored at or near the bottom, even as our college and university system was, and is, the best in the world. Incorrectly believing American students used to excel hampers our reform efforts because it makes the challenge of improving STEM education seem easier than it is.[54]

Misconception 2: American Students Don't Have the Aptitude for Math and Science

Drew wrote, "Fact: when American students struggle, teachers and parents attribute this failure to low aptitude. When students in many top-ranked nations, like Japan, don't perform well, teachers and parents conclude students have not worked hard enough. Aptitude has been overrated as a factor in achievement."[55]

Misconception 3: America Must Reform Its Math and Science Curriculum

Drew wrote, "Fact: Ever since *Sputnik*, the federal government has invested huge sums in curriculum reform, aka 'new math,' where the idea was to teach schoolchildren theoretical math concepts. But the students didn't get it, parents couldn't help, and new math was about as successful as 'new Coke.' Better that your child be taught by an exciting teacher using outdated text than by a boring teacher using the latest curriculum."[56]

Misconception 4: We Need to Recruit the Best and Brightest to Teach

Drew wrote, "Fact: the problem is not recruiting people into teaching. . . . The problem is keeping them in teaching. Teachers work hard. They are not paid enough. They endure great stress daily. 50 percent of teachers leave the profession within five years of entry. It is like pouring water into a sieve. We must develop and implement effective strategies for retaining talented teachers."[57]

Misconception 5: Only the Top Students Should Become Math and Science Teachers

Drew wrote, "Fact: excellent teaching requires more than simply possessing knowledge. You have to know how to communicate this knowledge. We need talented teachers who thoroughly understand the subject matter."[58]

The Long and Winding Road

So there you have the road map to the U.S. math and science skills gap: 49 years of reports and test results; data from many stakeholders of the American economy—from associations, trade groups, international and domestic testing services, government officials, top-notch business executives, respected consulting firms, and prestigious universities.

The data seem to point toward a single conclusion: For a very long time, the American education system has failed our country. For a very long time, the methods our country uses to teach our children have failed to produce results. For a very long time, the U.S. mathematics and science technology skills gap has been widening.

The question I have after researching and presenting all this data is this: How long will it be until the United States reaches the tipping point, the point at which the science and math skills of the rest of the world are just so far ahead of ours that America's economic, workforce, and national security are at risk? Our country's future hangs in the balance.

In Part Three, we will learn about the incredible efforts by companies and IT executives to help fill America's math and science skills gaps with "human mortar" made of ingenuity and innovation.

Notes

1. Toba Beta, *My Ancestor Was an Ancient Astronaut* (Indonesia: Primary Launch, 2011).
2. National Academy of Sciences, *Rising above the Gathering Storm Revisited: Rapidly Approaching Category 5* (Washington, DC: National Academies Press, 2010), www.bradley.edu/dotAsset/187205.pdf.
3. Ibid.
4. Ibid.
5. NASA Quest, Female Frontiers, "Meet: Jean J. Bartik, First Programmers Started Out as Computers," http://quest.arc.nasa.gov/space/frontiers/bartik.html.
6. Christianne Corbett, Catherine Hill, and Andresse St. Rose, *Why So Few Women in Science, Technology, Engineering, and Mathematics?* (Washington, DC: American Association of University Women, 2010), 14.
7. Ibid., 15.
8. Ibid., 16.
9. David Kaplan, "How *Waiting for Superman* Misses," *Fortune*, October 28, 2010.
10. Program on Education Policy and Governance Survey, 2010, *Education Next*, http://educationnext.org/files/Complete_Survey_Results_2010.pdf.
11. Craig R. Barrett, "Setting Standards in Education," Bloomberg Television, October 14, 2010, www.youtube.com/watch?v=DFcdpPy_C8k.
12. Quoted in Byron Auguste, Paul Kiln, and Matt Miller, *Closing the Talent Gap: Attracting and Retaining Top-Third Graduates to Careers in Teaching*, McKinsey & Company, September 2010, www.mckinseyonsociety.com/closing-the-talent-gap.
13. Ibid.
14. Ibid.
15. Ibid.
16. Ibid.
17. Ibid.
18. Ibid.
19. Ibid.
20. Ibid.
21. National Assessment of Educational Progress, Washington, DC, 2005, http://nces.ed.gov/nationsreportcard; National Assessment of Educational Progress, 2011.
22. Intel Corporation, "Exposure to Engineering Doubles Teens Career Interest," December 6, 2011, http://newsroom.intel.com/community/intel_newsroom/blog/2011/12/06/exposure-to-engineering-doubles-teens-career-interest.
23. Quoted in Eric H. Hanushek, Carlos X. Lastra-Anadon, Paul E. Peterson, and Ludger Woessmann, *Globally Challenged: Are U.S. Students Ready to Compete?* (Cambridge, MA: Harvard University Kennedy School of Government, 2011), iv, www.hks.harvard.edu/pepg/PDF/Papers/PEPG11-03_GloballyChallenged.pdf.
24. Ibid.
25. Ibid.
26. Ibid.
27. Thomas Loveless, *How Well Are American Students Learning?* (Houston: Brookings Institute, 2012), www.brookings.edu/~/media/newsletters/0216_brown_education_loveless.pdf.
28. Jason L. Riley, "Was the $5 Billion Worth It?" *Wall Street Journal*, July 23, 2011.
29. Loveless, *How Well Are American Students Learning?*
30. Ibid.
31. Ibid.
32. Ibid.

33. Council on Foreign Affairs, *U.S. Education and National Security* (Washington, DC: Council on Foreign Affairs, 2012), www.cfr.org/united-states/us-education-reform-national-security/p27618.
34. Ibid.
35. Ibid.
36. Ibid.
37. Michael E. Porter and Jan W. Rivkin, *Prosperity at Risk: Findings of Harvard Business School's Survey on U.S. Competitiveness* (Cambridge, MA: Harvard Business School, 2012), www.hbs.edu/competitiveness/pdf/hbscompsurvey.pdf.
38. Ibid.
39. Ibid.
40. Ibid.
41. Ibid.
42. Ibid.
43. Ibid.
44. World Economic Forum, *Global Competitiveness Report, 2012–2013* (Geneva, Switzerland: World Economic Forum, 2012).
45. Ibid.
46. Elizabeth Craig, Charlene Hou, Smriti Mathur, and Robert J. Thomas, *Where Will All the STEM Talent Come From?* (Chicago: Accenture Institute for High Performance, 2012), 2.
47. Ibid., 5.
48. Ibid.
49. Ibid.
50. Janet Lorin, "SAT Reading, Writing Test Scores Drop to Lowest Levels, Bloomberg, September 24, 2012, www.bloomberg.com/news/2012-09-24/sat-reading-writing-test-scores-drop-to-lowest-levels.html.
51. Institute of Education Sciences, National Center for Education Statistics, "Fast Facts," http://nces.ed.gov/fastfacts/display.asp?id=171.
52. Gaston Caperton, "SAT Report: Only 43% of 2012 College Bound Seniors Are College Ready," College Entrance Examination Board, press release, September 24, 2012.
53. American College Testing (ACT), *The Condition of College and Career Readiness, 2012* (Iowa City: ACT, 2012).
54. David E. Drew, "The Five Misconceptions about Teaching Math and Science," *Slate*, June 19, 2012, www.slate.com/articles/technology/future_tense/2012/06/science_education_myths_that_keep_us_from_fixing_the_system_.html.
55. Ibid.
56. Ibid.
57. Ibid.
58. Ibid.

PART THREE

Let's Build
Some Arks

A sk your colleagues this question: Who was the best businessperson in the latter half of the twentieth century? Many will say Jack Welch, the former General Electric chairman and CEO.

But I wouldn't. My choice for that position of honor is Louis V. Gerstner Jr., the brilliant executive who joined the IBM Corporation in 1993 just as pundits were claiming that the company was failing and should be split up and sold off. Gerstner, however, rejected that advice, kept the company together, and returned it to profitability.

He also had a keen sense for the future.

In 1995, more than 12 years before smartphones and tablets became best-selling devices that surpassed the sales of personal computers, Gerstner predicted the demise of the personal computer. And four years after that, he proclaimed at a Wall Street analyst meeting that overpriced Internet companies were "fireflies before the storm."[1] That was one year before the dot-com bubble burst.

A thoughtful manager—and a customer of IBM when he was chairman and CEO of RJR Nabisco—Gerstner often asked his fellow IBM workers for advice on what the company needed to do. A favorite Gerstner saying was "Execution is the tough, difficult daily grind of making sure the machine moves forward, meter by meter. Most important, no credit can be given for predicting the rain, only for building arks."[2]

Gerstner wanted solutions, not more problems.

His ark comment fits this section of the book. In the first two parts I have shared stories, report warnings, and test results I discovered in my five years of research. That research also introduced me to impressive work being done by private firms and nonprofit organizations to support science and math education in America.

Some might say, "How can you claim in an earlier section that no one was listening to all the warnings, and then recap the work of scores of firms pitching in to help?"

That's a valid comment, to a point. But even though I am going to present examples of private industry and nonprofit contributions to improving STEM education that date back to 1965, most of these innovative efforts began only in 2000. From 1965 to 2000, industry mostly sat on the STEM sidelines.

My primary goal in presenting these examples of great work, divided into three chapters focused on specific periods, is not to critique when they started but rather to inspire you to get involved in some way: by starting a STEM project at your company or in your community or by contributing your time and energy to one of the many organizations about to be profiled.

There's lots of work to be done.

Notes

1. "Riding the Storm," *Economist*, November 4, 1999.
2. Louis V. Gerstner Jr., *Who Says Elephants Can't Dance?* (New York: Harper Business, 2002), 231.

CHAPTER 13

Patchworking the Tech Skills Gap Begins

If you work hard at it, you can grind even an iron rod down to a needle.

—CHINESE PROVERB

1965: Skills USA

Founded as the Vocational Industrial Clubs of America in 1965 (the same year that the landmark Elementary and Secondary Education Act was passed), Skills USA, the name adopted in 1998, is "a partnership of students, teachers, and industry representatives serving teachers and high school and college students who are preparing for careers in technical, skilled, and service occupations."[1] The organization annually serves more than 300,000 students in local chapters across the country and welcomes volunteers and mentors from the IT industry.

With much of American education in the past 50 years focused on getting a college education, public interest in career and technical education (CTE) often gets brushed aside, with only one-third of parents saying they would "encourage their kids to work in a trade."[2]

That's a serious mistake.

According to a September 2012 report on CTE from Georgetown University's Public Policy Institute, by 2020 two out of three U.S. jobs will require some post–secondary school education and training (see Table 13.1).[3]

Another way to interpret the Georgetown University data is this: by 2020, 66 percent of all jobs in America will *not* require a bachelor's, or higher, degree; this represents a significant decrease from 1970, when 84 percent of U.S. jobs did not require a bachelor's, or higher, degree.

Table 13.1   Degrees Required for Jobs

Highest Level of School	1973	2010	2020
Master's degree or better	7%	11%	11%
Bachelor's degree	9%	21%	24%
Associate degree	12%	10%	12%
Some college but no degree	—	17%	18%
High school diploma	40%	30%	24%
Less than high school	32%	11%	12%

According to Skills USA, (1) 10 million new skilled workers will be needed in America by 2020, (2) 83 percent of companies report a moderate to serious shortage of skilled workers, (3) 69 percent expect the shortage to grow in the next five years, and (4) 600,000 skilled jobs are unfilled right now.[4]

A STEM initiative that tech workers might find interesting is the Engineering Alliance, a Skills USA partnership with Project Lead the Way and the Technology Students Association. The Engineering Alliance is an annual competition that "gives students the opportunity to showcase their STEM and leadership skills."[5]

Do your homework to learn more about CTE. It is a critically important and often overlooked component of the U.S. education system. Log on to www.skillsusa.org to learn how you and your company can get involved in a local Skills USA chapter.

1968: The Xerox Science Consultant Program

The Xerox Corporation is a rather remarkable company. Although most IT and business executives in 2013 are familiar with the company's high-profile position in the document management, digital color printing, and business services markets, fewer are most likely aware of the firm's amazing list of important contributions to the IT business.

This list includes the personal computer, the local area network, the mouse, the laser printer, and the window-partitioning software. Each was invented at the company's Palo Alto Research Center in the late 1960s or early 1970s.

However, in 1968, in Rochester, New York, then the company's headquarters (and one of the more innovative tech cities in America at the time), Xerox "invented" something else: it introduced the Xerox Science Consultant Program, which, according to my research, is the first corporate STEM initiative in the United States.

Although the Science Consultant Program started as a partnership with the American Chemical Society and the Rochester Council for Scientific Societies, Xerox took on full sponsorship in 1971 and has managed the program ever since.

The primary goals of the Xerox Science Consultant Program are really quite simple: (1) early engagement with students in the third to sixth grades, (2) reinforcement of a child's natural curiosity and interest in science and technology, (3) hands-on classroom activities that make acquiring knowledge fun, and (4) positive role models of STEM workers for students.[6]

For the past 45 years, Xerox engineers and scientists have matched their skills to 72 third- to sixth-grade elementary classrooms in Rochester and Webster, New York. Moreover, Xerox continues to generously supply materials for the science kits that the students need for hands-on practice of basic scientific principles.

Here's how the program works. Xerox scientists, engineers, and other technical workers partner with the Rochester City School District to develop, and help teachers present, science lessons in biology, chemistry, earth science, and physics for third- to sixth-grade students.

Xerox employees receive management support and are granted time away from work to prepare and conduct the lessons. About 100 Xerox engineers and scientists volunteer classroom instruction time twice a month. Throughout a school year, Xerox employees lead about 16,000 practical experiments with Rochester's students. That's about 720,000 experiments since 1968.

Lou Bostic, a senior technology executive at Xerox, observed that "the science kits [and experiments] are key to bringing the magic of science to the kids. Through the experiments they really get the science concepts because it is a way they can see, feel, and hear it."[7]

School officials are elated with the results of the program. One noted that "the practical simplicity that the Xerox consultants use to demonstrate science concepts works to reinforce the lesson that the student is learning in the classroom. The partnership benefits not only our students but our teachers as well."[8]

Mary Thomas, the principal of School 16 in Rochester, said the following:

I have been involved with the Xerox Science Consultant Program for the past 20 years. If my children are showing weaknesses in a certain area—for example, electrical energy—I know we can support the weaknesses with traditional education resources like textbooks, videos, etc. We also need resources to conduct the hands-on experiments. That is where Xerox comes in. We really believe we are better able to address the learning styles of all of our students with the support and resources supplied by Xerox.

And the program is delivering a real return on investment. Thomas added, "The Xerox Science Consultant Program and people are also key to boosting test results. The program has had a direct effect on our New York State Elementary Science Program evaluation test scores."[9]

The Xerox Corporation, its management, and its employees, deserve credit for the vision to become the first company in the United States, in 1968, to support STEM education in the nation's classrooms. The business model of the Xerox Science Consultant Program is simple and replicable. Does your company have engineers, technical workers, and scientists? If you do, you have all the ingredients to start your own science consultant program with your local school district.

Xerox is particularly notable because the company has maintained its commitment to the Science Consultant Program for 45 years. Many companies launch or financially support a STEM cause for several years and then when there is a change in management, or market conditions weaken, the support disappears.

There have been seven CEOs at Xerox since C. Peter McColough launched the Science Consultant program in 1968. In the 1970s and 1980s, the company's imaging and copier business was under tremendous competitive pressure from lower-cost Japanese competitors. But 45 years later, under the current leadership of Ursula Burns, the company remains committed to the Science Consultant Program. That's a remarkable commitment.

You have read about how the Xerox Science Consultant Program continues to have a positive effect on schools in two upstate New York school districts. But there are 13,500 other school districts in our country—school districts that need STEM professionals like you to follow the lead of Xerox.

Although the Xerox Science Consultant Program utilizes experiments, the program itself is not an experiment. It works! Experiment with the model in your school district.

1989: Women in Technology International

Women in Technology International (WITI) is a respected organization focused on the advance of women in business and technology. Started in 1989 by Carolyn Leighton, WITI has 140,000 members, including 8 percent who are men. It is the first STEM initiative focused on women.

The mission of WITI is "to empower women worldwide to achieve unimagined possibilities and transformations through technology, leadership, and economic prosperity."[10] The go-to-market strategy of WITI is to provide a platform of connections and resources, which the organization primarily does through an array of events, career development workshops, news and information services, and research.

WITI's "Hall of Fame" members are women who have made "outstanding contributions to the scientific and technological communities to improve and evolve society." The list of past winners is a who's-who of female technology rock stars.

When groups like WITI are launched, I am always interested in the backstory, the seminal idea that led to the launch. Leighton describes her epiphany moment this way:

> I remember so vividly the day the idea for WITI was born. I was sitting in a coffee shop waiting for a colleague. As I waited I started reading the cover story of a business magazine which focused on why women were not making it to the top, reporting that there was a mere 2% increase of women into middle management positions during the previous 10-year period. It was a statistic I found startling, even though it supported the stream of incredible stories I was hearing from women in technology.[11]

Although that business magazine cover story sparked Leighton's interest, the more she thought about it, other issues, too, had led her to consider launching WITI in 1989:

> (1) e-mail, which was making it easier to communicate with large groups of women around the world, (2) women were growing increasingly frustrated as they realized that their success was not entirely dependent on competence and contribution, but on politics and the whim of the "ole boys network," (3) the media face of the feminist movement [in 1989] was represented by the National Organization [for] Women who unfortunately chose to present the face of angry, male-bashing women who were alienating the very men women needed on our side, (4) while at the time there were women's groups formed around singular disciplines—like math, science etc.—technology was creating the opportunity for a multidisciplinary direction, and (5) the women's groups I was part of were pretty disappointing, I felt my time was being wasted with personal complaints rather than strategic discussions around moving forward and succeeding.[12]

WITI was thus founded, and it continues to this day to be the premier networking association for women in technology.

If you are a female technology executive, or even a male one (remember that 8 percent of the members are men), and want to join, go to www.witi.com and look for the membership icon button on the home page.

1990: Teach for America

Wendy Kopp was a Princeton University senior in 1989 with strong convictions about inequality in the American school system. For her senior thesis she envisioned an organization called Teach for America that would selectively recruit college graduates to teach in the nation's high-risk, and poorly performing, urban and rural school districts.

One year after Kopp's graduation from Princeton, Teach for America became a reality, recruiting and placing 500 teachers in school districts across the country.

Here's how it works. The process to secure a teaching position in Teach for America is highly selective. Of the 48,500 applicants to the program in 2011, only 14 percent were admitted.[13] Once selected, Teach for America corps members attend a five-week program to prepare for their two-year teaching assignment. If a corps member is not a certified teacher (most aren't), they must take additional coursework during their orientation period to earn an alternative certification.

Although Teach for America is often criticized by school administrators and teacher union members who think the program's two-year model does not provide a long-term solution to America's urban and rural teaching challenges, the Teach for America alumni numbers tell an entirely different story. According to Melissa Moritz, the managing director of Teach for America's STEM Initiative, 63% of the volunteers continue their full-time careers in education, with 50 percent becoming full-time classroom teachers. Another 13 percent stay on as school administrators.[14]

Teach for America is increasing the number of corps members committed to STEM education. In 2006, the program recruited 630 volunteers to teach science and math to 50,000 students. By 2011, Teach for America had increased the number of STEM corps teachers by 260 percent with a recruitment of 1,640 science and math teachers now instructing 120,000 students across America.[15]

Want to change careers and join the Teach for America program? Opportunity abounds!

In the 2012 class of Teach for America volunteers, 23 percent were graduate students and professionals.[16] Teach for America actively uses four recruitment processes to find volunteers looking for second careers as STEM teachers:

1. Private sector recruitment agencies.
2. Reengagement with Teach for America alumni with STEM skill sets.
3. Organizational partnerships with groups like the American Chemical Society, the American Mathematical Society, and the Society of Hispanic Engineers.
4. Free and paid job boards like Science, Engineering Central, PhDs .org, and the science section on Craigslist.[17]

Teach for America just might be your ticket to your new dream job. Log on to www.teachforamerica.org for more details.

1994: Tech Corps

June 6, 1994, was a typical hot, steamy evening in Washington, D.C. As publisher of *Computerworld*, I was the master of ceremonies for the annual Computerworld and Smithsonian Institution's Search for New Heroes gala held in the National Building Museum, one of the most impressive buildings in the nation's capital. As my dinner conversation concluded with Larry Ellison, the CEO of Oracle, Gordon Eubanks, the CEO of Symantec Corporation (both award winners that evening), and Linda Roberts, the deputy under-secretary for IT at the Department of Education, a staffer tapped me on the shoulder. This was my signal that it was time to head backstage to prepare for my final duty of the evening, offering a toast to all the winners.

My closing remarks, scripted days before, would be read easily on two teleprompters on either side of the main podium. The only logistical issue I had to be concerned with was giving enough time for 45 white-jacketed wait-ers to pour champagne into 1,000 glasses as I got to the toast portion of the remarks. As I settled into a chair to review my comments, Glenda Cuttaback, the executive producer of the event, approached me with an urgent request.

"Gary," she said, "we have a problem. The waiters are having a difficult time opening the bottles of champagne backstage, so we need you to extend your closing comments two minutes or so before the toast to give us enough time to open the remaining bottles." Those who know me well realize that the extension of remarks is not a problem for me. I asked Glenda to delay my stage reappearance for several minutes as I thought of something more to say.

I had brought Noelle, my 12-year-old daughter, to Washington for a weekend of sightseeing before the Monday event. She had been impressed with President Kennedy's gravesite, and we had talked over the weekend about his accomplishments, including the formation of the Peace Corps.

The idea struck me: that would be my extension remark. I would chal-lenge the 1,000 men and women in the audience, all of them key IT execu-tives, to join me in starting a new organization called the Tech Corps. As best as I can recall, here's how I challenged the audience:

> Just as the Peace Corps, started thirty-three years ago, focused on recruiting volunteers who went into developing countries to help, among other things, build roads that tied those countries to the global economy, what if we, the leaders of the information technology business, this evening start a new effort? An effort called the Tech

Corps, and we volunteer our technology skills to help K–12 schools all across America make the digital connections to connect every classroom to the information superhighway [which is what many called the Internet in 1994].

The extension worked beautifully! I filled up the two minutes—my call for the start of the Tech Corps was met with silence—the waiters poured the champagne, and the event came to a conclusion.

Or so I thought. After a dessert reception I was standing in line to get a cab to leave the event when the person directly behind me tapped me on the shoulder and asked, "What is Tech Corps?" After I shared with him that I had just made up the idea that evening and hadn't thought it through, he said, "Well, I have, and I like it. Here's my card. What's your Social Security number? I want to meet you tomorrow morning at 9 a.m." My first thought was "Who is this guy?" but then I looked at his card. It had an embossed golden eagle in the middle of it and raised type that said THE EXECUTIVE OFFICE OF THE PRESIDENT OF THE UNITED STATES.

His name was Ed Fitzsimmons, and he worked in the Office of Science and Technology Policy in the White House. The next morning we did indeed meet, in a special room in the Old Executive Office Building at President Richard Nixon's desk (I even saw the special drawer where Rosemary Woods kept the hidden tape recorder). The game plan was to share the idea to a wider audience, implement a test concept, and, if all went well, formally launch Tech Corps as a nonprofit organization in the White House.

I did share the idea widely. One of my contacts, Harry Saal, the president of a network company, posted it on an electronic bulletin board, and within days, from all corners of the United States, I had thousands of responses from IT workers expressing their approval of the concept and asking, "Where do I sign up?"

Ed Fitzsimmons did his part, too. A big push forward for Tech Corps came when Fitzsimmons briefed William Curry, President Clinton's special counsel for domestic affairs, on the concept of Tech Corps. Curry liked it, took it to the president with his recommendation to support the initiative, and after a five-month test run in Massachusetts, on October 9, 1995—16 months after I had made the champagne extension comment—Tech Corps was launched as a national nonprofit organization in the Roosevelt Room of the White House with President Clinton and Vice President Gore jointly making the announcement.

Tech Corps was launched as an organization that would challenge American IT workers to volunteer their skills to help K–12 schools build their technology infrastructures. Back then, technology was a mess in public schools. Few schools had cohesive technology strategies, and the "newest" computer equipment was usually a donated machine that was several years old and still had old information on the hard drive that had to be erased. Tech Corps volunteers made sense of that tech chaos, and over the 18 years of its contributions to America's public and private schools, more than 10,000 Tech Corps volunteers have donated nearly $45 million worth of their consulting time.

As with any organization, the challenges and priorities have changed. In 2013, with all schools wired to the Internet and many students as tech-savvy as the early Tech Corps volunteers were in 1995, the goals of Tech Corps, too, have changed. Today the mission of Tech Corps is "to facilitate a technologically literate society in which *all* K–12 students have equal access to the technology skills, programs, and resources that will enrich their education and allow them to successfully compete in the global workforce."[18] It's a much higher goal than simply wiring computers and clearing hard drives!

My favorite Tech Corps program in 2013 is called Techie Club, a school year–long 40-hour program developed with support from the American Electric Power Foundation and aimed at third- to fifth-grade students. Techie Club programs focus exclusively on the T—technology—of STEM and expose students to different technology concepts to incite interest and encourage exploration in all things tech.

Techie Club programs are facilitated by IT volunteers who develop and assist Tech Corps staff in creating lessons and activities that encourage critical thinking, problem solving, communication, collaboration, and creativity—skills needed by students to succeed in the twenty-first-century global workforce. Techie Clubs are composed of up to 20 students and typically meet weekly throughout the school year for 60 to 90 minutes after school.

Wayne Hicks, a Techie Club volunteer, had this to say about his experience with third-grade students in Cincinnati, Ohio: "I watched them light up as they learned about bits and bytes, as they discovered component parts of a computer, as they worked in teams to create blogs, websites, and other creative projects. The Tech Corps Techie Club opened their minds to be creators and innovators of technology. The end of each volunteer session came much too soon."[19]

Frederick A. Borden, a software engineer with Cincom Systems Inc., shared with me why he offers his time as a Techie Club volunteer: "There is probably no better way to support our schools, teachers, and students than to invest a little time for a much bigger return, and still maintain your career. As a nation, we are falling behind in STEM education, and with fewer students entering these fields of study, this is problematic for a knowledge-based economy, and it has to change."[20]

Hicks offered up this challenge to the readers of this book: "I truly recommend that all IT professionals reading (name of book) take a moment to assess their impact on our nation's efforts to create a prepared workforce of the future. Ask yourself: 'Can I donate one hour per week to help unleash young minds in a positive direction?'"[21]

Readers should log on to www.techcorps.org to learn more about Tech Corps, its innovative programs, and how to start a Techie Club program in your area.

1995: NetDay

In April 1995, John Gage, the chief science officer (CSO) for Sun Microsystems, was attending a meeting of the Federal Networking Advisory Committee, which was discussing the fiscal challenges public schools faced in connecting their classrooms to the "information superhighway." The conventional wisdom at the time was that a network connection—including wiring, computers, routers, and labor—would cost a school about $2,000 per classroom.

Sitting in the meeting, Gage fiddled with some numbers and came up with another estimate: less than $100 per classroom. Part of Gage's calculations were based on pro bono contributions of wiring and labor from the industry, and when he flew home to California, he shared his idea with Michael Kaufman, a passionate technology activist and at the time a manager at the NPR affiliate in San Francisco.

Gage and Kaufman worked on their idea, which they called NetDay, during the summer, and that's when the all the fun began. You have to understand that Gage is one of the smartest, most convincing, and most affable technology executives in the industry. He's also a great salesman.

In August 1995, Gage arranged an impromptu meeting with Thomas Kalil, the special assistant to President Clinton for economic policy, where he laid out the idea for NetDay. He essentially described it as a one-day barnstorming effort in California to mobilize the tech industry to connect thousands of schools, libraries, and clinics to the Internet on Saturday, March 9, 1996. After the meeting, Gage quickly built a mock web page of the NetDay site and sent it along to Kalil.

Although NetDay was still just an idea, Gage and his mock NetDay web page were so convincing that on September 21, 1995, at a tech event held at the Exploratorium in San Francisco, President Clinton—surrounded on stage with a group of rock-star tech CEOs—proclaimed, "I came here to San Francisco today to issue a challenge to America to see to it that every classroom in our country is connected to the Information Superhighway. By the end of this school year, every school in California, all 12,000 of them, will have access to the Internet and its vast world of knowledge."[22]

The speech was a huge success, and the following day the *San Francisco Chronicle* touted it in a front-page article with the headline DOUBLE CLICK ON CLINTON.

Six months later, after an incredible amount of work by Gage and Kaufman, NetDay became a reality when President Clinton, Vice President Gore, and volunteers around the state, including myself, connected thousands of schools and classrooms to the Internet.

The morale of NetDay is if you have a good idea, push it hard!

1996: SAS Curriculum Pathways

James Goodnight, the chairman of the SAS Institute—a Cary, North Carolina, data analytics firm he founded in 1976—is on my personal short-list of the business executives most committed to STEM education in America.

In fact, he and his wife, Ann, were so committed to STEM education—and, frankly, frustrated with the local public school system—that in 1996 they contributed $10 million of their own money and, along with John and Ginger Sall, created the Cary Academy, an independent college prep school for grades 6–12 whose mission was to be a model school for integrating technology into all facets of education.

The Cary Academy, sometimes called Goodnight High, is a remarkable place. You, too, can "visit" it by logging on to www.caryacademy.org.

But perhaps even more impressive than the Cary Academy, which enrolls about 800 students, is the SAS Curriculum Pathways program, also launched in 1996 as SAS in Schools, the first corporate-supported online learning initiative in America, now used by 50,000 teachers and 12,000 schools across the United States.

Available to educators at no cost, SAS Curriculum Pathways is an interactive, online initiative aimed at improving the math and science (and other subjects) skills of students in grades 6–12. SAS Curriculum Pathways is an excellent example of how technology can be leveraged to teach more complex cognitive, collaborative, critical thinking skills, which are needed by today's students.

George Ward, the lead teacher at Centennial Campus Middle School in Raleigh, North Carolina, said that the SAS Curriculum Pathways program "has allowed me to become more effective as a teacher because I can diagnose what my students are learning and what they aren't. It's become more of a student-centered classroom."[23] Centennial is participating with the SAS Curriculum Pathways program and the 1:1 Laptop Initiative in a project in which students "don't do much with pencils and paper anymore and complete lessons, create multimedia projects, and take tests on school-issued laptops using SAS Curriculum Pathways software."[24]

SAS Curriculum Pathways caught my attention for another reason. It is positioned to be the next critical transformation shift in public education in America: the shift from print to digital content and the adoption of personalized content. Scott McQuiggan, the director of SAS Curriculum Pathway, addresses that transformation this way: "As the transformation from print to digital content takes hold, we believe that SAS Curriculum Pathways content will not only help schools in their efforts to incorporate digital curriculum, but also better enable teachers to personalize lessons to the individual needs of every student."[25]

The Cary Academy and SAS Curriculum Pathways are excellent examples of James Goodnight's vision to show America how to incorporate technology into education. In an interview with *Forbes*, he said he launched the Cary Academy and the SAS Curriculum Pathways program because "I wanted to create a place for learning where teachers could be mentors, where the technology is right at their desks. The teacher sees if students are struggling or if they are playing solitaire."[26]

Why is James Goodnight so committed to STEM education? He explains it with these words:

In this information age, challenges that require STEM skills—science, technology, engineering, and mathematics—will only increase in the years to come. If American students aren't equipped to do the work, there are tens of millions of people in Asia who will step in and take those jobs. And the next generation of American workers will become service workers. Today's kids are born with a cell phone in one hand and an iPod in the other. They text-message each other, use Facebook, YouTube, and Google. They play games on their Nintendos, PlayStations, Xboxes, and the web. Yet when they come to school, they enter a classroom with a teacher and a blackboard. Education hasn't changed in 100 years. Kids are bored. How do you make school interesting again? You give them tools they like to use.[27]

For the 2012–2013 school year, the Cary Academy has widely distributed tablet computers to students.

In this era of strong focus on testing as required by the No Child Left Behind Act, the SAS Curriculum Pathways program is delivering results. After a year of experiencing the program at the Centennial Campus Middle School, eighth-grade students have achieved high growth in their state math tests as calculated by the North Carolina ABCs accountability measurements, and the school's passing rate on the state-mandated computer test has increased by 16 points.[28]

SAS Curriculum Pathways is available for any school to use by logging on to www.sascurriculumpathways.com. Review the curriculum materials and then set up a meeting with the heads of your school's science and math departments to discuss how you can integrate SAS Curriculum Pathways into your district.

1997: The Cisco Networking Academy

Several years after the high-profile August 1995 initial public offering of Netscape, an event that launched the evolution of the letter *e* from a vowel into an adjective, as every business was creating "e-this" and "e-that" new offerings, K–12 schools were largely still sitting on the "e-sidelines."

However, with the 1997 congressional passage of E-Rate—also known as the Schools and Libraries Program of the Universal Service Fund—telecommunications carriers were obliged to collect a surcharge on every business and residential phone bill to fund the project of connecting schools, libraries, and rural areas of the United States to the Internet.

Schools were elated by the program and began to buy network switches, routers, patch cables, and RJ-45 jacks to build those connections. John Morgridge, then president of Cisco Systems, said his company became interested in the school market when "one of our employees actually climbed a fence dividing our headquarters and a public school next door to Cisco and gave the school a router." Another employee, according to Morgridge, drove to a nearby Catholic school to donate a router to it, and the principal, who was also a priest, said to the employee, "I don't know what a router is, son, but I'll take one."[29]

The funds from the Federal Communications Commission's E-Rate program created a huge opportunity for network equipment providers like Cisco, which embraced the market with an initial program called Take a Router to School. Morgridge claims that Cisco Systems "shipped a lot of routers to schools but there was a big problem when the routers arrived at those schools: they didn't have the tech staff who knew how to effectively use a router. So Cisco's Networking Academy program grew out of that need."[30]

With the e-commerce market growing rapidly at the time, the demand (and the salaries) for skilled network engineers and tech staff was huge. With fewer fiscal resources, schools couldn't compete.

That's when an enterprising Cisco Systems employee actually came up with the idea that evolved into the Cisco Networking Academy. George Ward's first attempt was to develop a special network curriculum focused on the needs of K–12 IT users. But with school IT professional efforts spread among integrating a plethora of new computers into an infrastructure of old machines and writing software programs for those computers, there was not enough time to learn the new market of network routers and switches.

Ward had a plan B. He reached out to students at the middle schools and high schools where Cisco Systems had sold its routers and switches and developed a special network curriculum for students, rather than IT professionals, in which students could be trained during after-school hours to manage the school's network.

Ward's innovative student-based program caught the imagination and interest of school administrators and workforce development groups around the country, and they asked Ward if he could formalize and scale the curriculum program. Ward responded to the challenge by writing a simple network curriculum that could be taught as elective courses in a four-semester program. The curriculum focused on the principles and practices of designing, building, and maintaining a network, with a big reward for students upon completion of the course and passing an exam: the opportunity to become a Cisco certified networking associate.

The Cisco Networking Academy was an instant win-win. K–12 schools won because they now had inexpensive, skilled network professionals who could connect the school's classrooms to the Internet. And students enrolled in a Cisco Networking Academy program had a jump-start on a high-paying future job. Cisco Systems won because it was training an army of network technicians skilled in the company's routers and switches.

Since its launch 16 years ago, the Cisco Networking Academy program has spread to 10,000 academies in 165 countries and has trained 4 million students. Joy Miller, a Networking Academy graduate in Pennsylvania, was introduced to the program by a Pennsylvania workforce development organization. Here's what she had to say about what the Cisco Networking Academy has meant to her life:

> I was a stay-at-home mom for ten years, and when my husband and I divorced I was on public assistance and couldn't imagine what kind of job I could get with three kids, no home, and no skills. The Cisco Networking Academy gave me a career—not just a job. I started at Toll Brothers and now work as a network engineer at LabCorp Inc., where my starting salary was in the high five-figure range. If you would've told me three years ago that I'd be bringing home this kind of pay I'd have said, "How? How could I do that?"[31]

So what exactly is a Networking Academy? Is it a physical building? Can I find one on Google Maps? How can I participate? Log on to www.netacad. com. The site has all the information on how you can join a Networking Academy as a volunteer instructor or start one yourself.

1998: I.C.Stars

Sandee Kastrul began her professional career as a math and science teacher in Chicago, where a favorite part of her job was continuing her relationship with her former students. "When my kids would come back to visit me, we would have great conversations," said Kastrul. "I would ask them things like 'What are you doing with science?'"[32]

One day, however, a conversation with one of her most talented former students changed her life. When she asked this student her "What are you doing with science?" question, the student answered, "I'm working with chemicals." Kastrul said she got all excited and asked, "What are you doing?" When he responded, "I am working in cleaning services in one of the hotels downtown, and you would be surprised that people don't know about the basic properties of ammonia," Kastrul was "heartbroken."[33]

It inspired Kastrul to collaborate with her friend Leslie Beller to launch I.C.Stars as a workforce development nonprofit organization in 1998.

When they launched I.C.Stars, Kastrul described her ideas as follows:

> There was a big shift happening in business, the world was becoming smaller and larger at the same time. Our competitors weren't necessarily the companies down the street but around the globe. The workers of the next century would need to be people who thought out-of-the-box, who could figure out different rules of culture and had an innovative mindset. So many of the young people I had worked with in the classroom possessed those skills, not because they had gone to school to learn them, but because as they had gone through life they had to learn them. We saw a real path in the concept that if American businesses were going to thrive in the next century, they would need to employ people who came from this school of wisdom. There was a great need for young people to be able to plug into a system that valued their contributions and could see the leadership, the resiliency, and the reciprocity that they possessed as a total leadership concept.[34]

The highly selective I.C.Stars program targets young adults (18–27 years old) who often do not have a formal education but who, according to I.C.Stars, "actually represent an untapped source for future economic and social leadership."[35] Using a project-based learning module, I.C.Stars participants first must complete a 16-week internship. Once that is completed, they enter a two-year program focused on technology training and leadership development while continuing to earn an associate's degree from a city college in Chicago.

Ninety-five percent of the individuals who enter the program had average yearly earnings of $9,000 before they started the I.C.Stars program. After graduating from I.C.Stars, on average, they earn $31,000.[36]

My favorite I.C.Stars initiative is an innovative program called High Tea that matches CIOs with young adults. High Tea is a highly popular volunteer-mentoring opportunity for CIOs in Chicago. From 4 to 5 p.m.

every day during the 16-week internship program, a Chicago-based CIO visits the offices of I.C.Stars to talk with the interns about their journey to success, impressing upon the aspiring interns the importance of perseverance and setting short- and long-term goals. Kastrul said that "this inspirational hour is one of the greatest strengths of our training program."[37]

That's 90 CIOs who participate in each 16-week internship session. And there is a waiting list among IT executives to participate, according to Kastrul.

Dave Edelstein, the CIO of Siemens, is a huge supporter of High Tea. On the I.C.Stars web site, Edelstein wrote the following:

> I discovered I.C.Stars over 10 years ago as I was searching for sources of candidates from diverse backgrounds. I attended a tea and not only did I find a source of candidates from diverse backgrounds, I also found a set of young people who are bright, energetic, hardworking and eager to get a chance to succeed. And the I.C.Stars interns that I hired not only excelled at their given tasks but infused everyone, even SAP veterans, with new levels of energy and enthusiasm.[38]

Tiffany Mikell, a graduate of the I.C.Stars twelfth cycle (class), joined the program shortly after giving birth to her son:

> Even though I didn't have any technology skills I was a logical thinker and knew I needed a career—and I needed to be a positive role model and provide a good future for my son. I enrolled in I.C.Stars, and even though the rigorous training program concerned me at first, I stuck with the program, which helped me overcome personal obstacles and become an adept problem solver. After graduation from I.C.Stars I enrolled in a 30-day Java boot camp with a few dozen other young people, hoping to get hired by Accenture. Many of the other participants had a college degree, but the skills I acquired through the I.C.Stars project-based curriculum allowed me to compete and get hired by Accenture, where, four years later, I work as an associate software manager and volunteer time to help other young people in the community.[39]

If you are a CIO working in the Chicago area, get involved with I.C.Stars and High Tea. If you live in other parts of the country, I.C.Stars and High Tea are noteworthy examples of the ways that IT executives can reach out to less privileged youth to share with them the wonderful opportunities in the IT business.

All you have to do is use your imagination.

1998: Intel Teach

In 1998, the Intel Corporation launched Intel Teach. Now in its fifteenth year, Intel Teach has trained 10 million teachers around the world.

In the late 1980s and throughout the 1990s, business leaders identified a critical skills shift that was occurring in their workforces: a shift away from manufacturing skills to cognitive skills. Front and center was the need for workers who thought differently and worked differently. The Intel Corporation responded to this need with Intel Teach, "a professional development program that enables educators to effectively integrate technology into their lessons to promote problem solving, critical thinking, and collaboration skills among students."[40]

Six of the courses in the Intel Teach professional development series that address these core twenty-first-century skills in an e-learning format are the following:

1. **Critical thinking with data.** This examines critical thinking in data analysis.
2. **Assessment in twenty-first-century classrooms.** This helps students to plan, develop, and manage student-centered assessment strategies.
3. **Project-based approaches.** This explores project-based learning with classroom scenarios to facilitate student-directed learning.
4. **Collaboration in the digital classroom.** This helps students to plan and manage collaborative activities, with an emphasis on online tools
5. **Blended learning.** This helps teachers and students to learn in both the traditional classroom and online.
6. **Inquiry in science.** This promotes the transition from textbook science to authentic inquiry, thinking, and practice.[41]

I asked several Intel Teach educators how the program works in their classroom.

Glen Westbroek, a middle school teacher in Lindon, Utah, who is an Intel Teach teacher participant, told me the following:

Each 21st century skill (collaboration, critical thinking, leadership) presents its own challenges to me as a teacher. Students often think that collaboration means one person does the work so others may copy it. Intel Teach helps me help students understand that true collaboration means discussing and coming to agreement on concepts. Critical thinking, however, is the most challenging skill to teach because the 7th grade students I work with are used to being told "what to memorize." Intel Teach helps my students begin to feel they can think more deeply after they use these tools.[42]

A core component of Intel Teach is its focus on project-based experiences rather than lecture and memorization. That's important to Angie Hillman, an Intel Teach master at Cottonwood Middle School in Arizona, who said, "In this day and age, with all the state testing and other things weighing on educators, I had lost my focus on how to make learning fun. In creating my Intel unit, I have been able to bring fun back into my classroom with the highest level of instruction possible for my students."[43]

I also asked Intel Teach educators, "How do you measure progress in teaching collaboration, critical thinking, and leadership?"

Brant Benniga, a high school social studies teacher in Maize, Kansas, measures progress in critical thinking this way: "One measure I use is how students respond to new activities compared to earlier activities. When I see them tackle a new task with fewer questions and more confidence, I feel like progress is being made."[44]

Stacey Ryan, a sixth-grade math teacher in Andover, Kansas, has a different approach to measuring the progress her students are making with Intel Teach: "When all students feel the confidence to explain their ideas, argue their opinion, and express confusion in a clear and thoughtful way, I know I am making headway."[45]

Intel Teach can help the teachers in *your* school district teach *your* children the skills they need to learn to compete in the twenty-first-century workforce. To learn more, log on to http://intel.com/content/www/us/en/education/k12/intel-teach-us.html.

Notes

1. Skills USA, "Skills USA Fact Sheet," n.d., www.skillsusa.org.
2. Skills USA, "America's Skills Gap," n.d., www.skillsusa.org.
3. Anthony P. Carnevale, Andrew R. Hudson, and Tamara Jauasundrea, *Career and Technical Education: Five Ways That Pay* (Washington, DC: Georgetown University Public Policy Institute, 2012), 2.
4. Skills USA, "America's Skills Gap."
5. Lisa Cohen, "Contests Add More Excitement," Engineering Alliance, March 2, 2010, www.prweb.com/releases/engineering_competition/STEM_education/prweb3660044.htm.
6. Ewart LeBlanc, "Can Corporate America Help? How Xerox Does It with the Science Consultant Program," American Association of Physics Teachers, April 16, 2004, www.powershow.com/view/1791e-ODRkN/Can_Corporate_America_Help_powerpoint_ppt_presentation.
7. Xerox Corporation, "Xerox People Bring the Magic of Science to Schools," n.d., http://xerox.com/about-xerox/citizenship/news/science-consultant/enus.html.

8. Xerox Corporation, "Xerox Science Consultant Program: Bringing Science to Life for 40 Years," press release, October 30, 2007, www.happynews.com/news/10302007/xerox-science-consultant-program-bringing-science-life.htm.

9. Ibid.

10. Women in Technology International, "About WITI," n.d., www.witi.com.

11. Ibid.

12. Ibid.

13. Alison Damast, "Q&A: Teach for America's Wendy Kopp," *Bloomberg Business Week*, March 26, 2012.

14. Melissa Gregson, e-mail to author, September 21, 2012.

15. Teach for America, "Math and Science Education Initiative," n.d., www.teachforamerica.org.

16. "Recruiting STEM Professionals," attachment to Gregson e-mail.

17. Ibid.

18. Tech Corps, "About Us," n.d., www.techcorps.org.

19. Wayne Hicks, e-mail to author, September 21, 2012.

20. Frederick A. Borden, e-mail to author, September 12, 2012.

21. Hicks e-mail.

22. President William Jefferson Clinton, "Remarks in the Exploratorium," San Francisco, September 21, 1995.

23. SAS Institute, "SAS Curriculum Pathways Engages Students," n.d., www.sas.com.

24. Ibid.

25. SAS Institute, "Back to School—SAS Curriculum Pathways," August 15, 2012, www.sas.com.

26. Stephanie Cooperman, "Goodnight High," *Forbes*, April 21, 2008.

27. Ibid.

28. SAS Institute, "Back to School."

29. Cisco Systems, "Cisco's 25th Anniversary Video Blog Series," December 2009, www.youtube.com/watch?v=Nu7FhntLv4U.

30. Ibid.

31. Cisco Systems, "Mind Wide Open," Cisco Networking Academy, 2007, www.netacad.com.

32. I.C.Stars, "A Star Is Born," n.d., www.icstars.org.

33. Ibid.

34. Ibid.

35. Ibid.

36. I.C.Stars, "A Measurable Impact," n.d., www.icstars.org.

37. I.C.Stars, "High Tea," n.d., www.icstars.org.

38. Ibid.

39. I.C.Stars, *2010 I.C.Stars Annual Report*, 2011, www.icstars.org.

40. Intel Corporation, "Intel Teach Elements—Online Professional Development Courses," n.d., www.intel.com/content/www/us/en/education/k12/teach-elements.

41. Ibid.

42. Glen Westbroek, e-mail to author, September 2012.

43. Intel Corporation, "Intel Celebrates 10 Million Teachers Trained," September 2011, http://newsroom.intel.com/community/intel_newsroom/blog/2011/09.

44. Brant Benniga, e-mail to author, September, 2012.

45. Stacey Ryan, e-mail to author, September, 2012.

CHAPTER **14**

The Pace of Remediation Work on the National Skills Gap Accelerates

If you want something done, ask a busy person.
—BENJAMIN FRANKLIN, *POOR RICHARD'S ALMANACK*

2000: Year Up

I first met Gerald Chertavian, the founder of Year Up, during the spring of 2005 at a Massachusetts Technology Leadership Council event in Newton, Massachusetts. Having founded Tech Corps in 1995, I was curious to learn more about Year Up. After I introduced myself to him, he said, "Please join me for the June graduation luncheon honoring the class of 2005 Year Up students."

Several weeks later I was sitting in a banquet room at the Sheraton Hotel in downtown Boston. There were maybe 100 people in attendance honoring the approximately 20 graduates of the yearlong (hence the name Year Up) program. I left impressed, and several weeks later, over lunch, Chertavian told me the story of how he started Year Up.

A 1987 graduate of Bowdoin College, Chertavian was living in New York City and joined the Big Brother volunteer organization, where he met 10-year-old David Heredia. For three years, Chertavian spent every Saturday with David and realized that David lacked the resources for the path to success that Chertavian had access to.

After his Big Brother experience with David, Chertavian headed off to Harvard Business School, where after graduation he cofounded and later sold a dot-com company. But Chertavian never forgot his experience with David Heredia. So in 2000 he decided to start Year Up to help close the digital divide.

After that lunch I kept in touch with Chertavian. He invited me to the 2006 Year Up graduation, but I could not attend because of a prior business commitment. I did attend the 2007 graduation, however, and that was my Year Up "wow" moment. The ceremony still took place at the Sheraton Hotel, but because the graduation class had grown to at least 140 young adults, the ceremony was moved to the hotel's main banquet room to accommodate more than 1,000 attendees! I was overwhelmed as all the graduates received their diplomas on stage; several were selected to tell their stories about how their experience with Year Up saved their lives.

The mission of Year Up is to close the digital divide by providing urban young adults with the critical technology, collaboration, leadership, and basic business skills they need to cross the chasm from unemployment to fulfilling professional careers. The organization, launched in Boston, currently has programs operating in Atlanta, San Francisco, Chicago, New York, Providence (Rhode Island), Seattle, and Washington, D.C. In 2012, there were 1,300 students enrolled in the program.

Year Up focuses on high school graduates or general education degree recipients ages 18–24. In an intensive 52-week program, Year Up addresses an individual's social, emotional, and technical development. While enrolled in the program, students receive a modest stipend and can earn college credits. Year Up is a highly selective organization, and it does no advertising. Prospective participants typically hear about the program from their friends. The mission of Year Up is to provide graduates a path to economic self-sufficiency.

The Year Up model works by attracting local sponsorship from major employers in the area served by a Year Up chapter. The employers work with Year Up to create a list of the workforce skills they most need. A Year Up semester consists of six months of technical skills training that is regularly updated to meet the needs of the sponsoring corporations. In addition to receiving basic tech training, Year Up students are also required to take classes in business writing and communication that focus on verbal communication, grammar, and composing and proofreading e-mails, memos, and reports.

A critical requirement of the Year Up program is that each student signs a performance contract that states that the student will maintain a high attendance rate, be on time, and complete the assignments in a timely manner. The program endeavors to instill six values in every participant: (1) respecting valuing others, (2) building trust and being honest, (3) being accountable, (4) engaging and embracing diversity, (5) working hard and having fun, and (6) always striving to learn.[1]

After the six-month technical and business skills development phase, each Year Up student is paired with a corporation in the area (usually one of the sponsoring corporations) to begin a six-month internship program to further hone the student's technology, business, and professional skills and to begin networking for a full-time job.

Since its inception in 2000, Year Up has successfully placed 100 percent of qualified students into internships. The overwhelming majority, 95 percent, of Year Up interns meet or exceed their internship managers' expectations, and 84 percent of Year Up graduates are employed (earning, on average, $30,000) or attending college full-time within four months after graduation.[2]

Chertavian succinctly frames Year Up's goal this way: "We want to change perceptions of urban young adults from societal deficits to societal assets."[3]

Several of those "assets" have shared their opinion of their experience on the Year Up website. Jonathan Garcia, a Year Up New York graduate, expressed his amazement with the Year Up program: "If you would have told me a couple of years ago that I would be interning at American Express, helping VPs, I would never in a million years have believed you."[4] Three thousand miles away, Hassan Anjum, a Year Up San Francisco Bay Area graduate, expressed his gratitude for his involvement with Year Up: "Having that door open for you and know that all you need to do is step through it is what Year Up provided me. Without that bridge, I would not have been able to do anything as far as professional careers go."[5]

Log on to www.yearup.org to learn more about this noteworthy nonprofit group. If you live in Atlanta, Boston, Chicago, New York, Providence, Seattle, San Francisco, or Washington, D.C., call the local Year Up chapter and volunteer your services as a tech instructor or mentor. If your city is not currently served by Year Up, call Year Up headquarters and speak with Gerald Chertavian about starting a chapter where you live.

2000: The Juniper Networks Foundation Fund

Four years after the company Juniper Networks was founded in 1996, it created the Juniper Networks Foundation Fund. I am impressed with the Juniper Networks Foundation Fund for two reasons: (1) the thoughtfulness of the founders in creating it so soon after the firm was launched, and (2) the innovative approach the company takes in managing this philanthropic effort.

Throughout this book you have read about many projects and organizations that are doing good work. Perhaps you might have even wondered how you could start something. The Juniper Networks Foundation Fund model is a good road map.

First, connect your foundation's goals directly to the business strategy of the company and the enhancement of the company's brand image. For Juniper Networks, one of the world's leading network companies, the strategic vision of the foundation was "connect and empower the unconnected." To implement the strategy you need tactical ideas. For Juniper, the signature themes were (1) empowering women and children, (2) connecting technology to nonprofit groups, and (3) using the new network in education, lessening the digital divide and solving hard problems.[6]

Next, after you have set the business strategy goals of your foundation, you need to determine several key components of the foundation: geographic and human resource scale issues. Juniper chose a global model because that approach best supported its business strategy, and it used business partners to help launch its foundation.

Regardless of which approach you choose, budgeting is critical. A key mistake many companies make when they start foundations is that they do not make the foundation's budget commensurate with the fiscal resources of the company. Obviously, if you spend too much, the foundation's model is not sustainable. Likewise, if you budget too little compared to your firm's ability to invest, it will not excite and inspire enough employees to get involved.

And that—the ability of a foundation to excite employees—is possibly the aspect of the Juniper Networks Foundation Fund that impressed me the most.

For nearly 20 years I have worked on boards of nonprofit organizations, and the most successful entities were those with committed boards. Juniper Networks approached board composition by going right to the most important stakeholders at Juniper Networks: the employees.

Because the Juniper Networks Foundation Fund is global, the fund is managed by 12 members from North America, Europe, the Middle East, Africa, and Asia. These 12 individuals are totally responsible for deciding which firms are awarded Juniper Foundation grants, they act as ambassadors for the foundation on a global scale, they are the first point of contact for grantee nonprofit groups, they are responsible for being keenly aware of new global initiatives the fund might invest in, and, most important, by their example they encourage other Juniper Networks employees to engage in volunteer efforts sponsored by the fund.

Another important component of the Juniper model is that its foundation's management committee has representatives from all key departments, all levels of management, and all levels of diversity. Each committee member serves a two-year term.

I wanted to know why members of the Juniper Networks Foundation Fund had decided to serve on the management committee. Stacey Clark Ohara, a Juniper Networks Foundation Fund strategy executive, shared a few members' answers with me: (1) "We are in a very well-paid industry. As my children grew up I decided the time was right to devote more time to something philanthropic, yet connected to my career"; (2) "Reviewing grant proposals has made me realize that there is a greater need for support than I originally thought—it has broadened my horizons"; and (3) "I can never give enough time to the committee. Reality always cuts off my plans. But increasingly I see the two things coming closer—the work of the Juniper Foundation feeds my daily role and vice versa."[7]

The readers of this book undoubtedly work in all types of companies: large and small, regional and global. Does your firm have an organized philanthropic foundation that strategically supports the business goal of the company, does good work, and gets employees involved and committed? My observation is that most companies do not, and those that do take a scattershot approach to philanthropic endeavors, spreading their efforts too thin among many very good organizations.

If you want to affect your business and do something good, the road map of the Juniper Networks Foundation Fund is all you need to follow to succeed.

2002: Technology Goddesses

In 2002, Cora Carmody, the CIO of Jacobs Engineering—a global engineering, architecture, and construction firm headquartered in Pasadena, California—was concerned about the low number of women enrolled in STEM undergraduate programs and the percent of women working in STEM careers.

But besides having a concern, she had a plan to address it. Carmody wanted to "make it cool for a young girl to be a geek," so she partnered with the San Diego regional chapter of the Girl Scouts office and launched Technology Goddesses to focus on "educating Girl Scouts of all ages in technology-related disciplines, with a component of service that brings the older girls into mentorship positions with elementary school–age girls, exposing the girls to the vast range and variety of technology careers."[8]

Before I learned about the Technology Goddesses program, I was unaware of any other STEM-related initiatives sponsored by the Girl Scouts, so I asked Carmody why she selected that group as a partner. "The Girl Scouts of the USA provide an excellent framework of achievement definitions at all ages," she said, "and they provide a solid infrastructure of girl and volunteer recruitment and a tremendous model of Girl-Adult Partnerships which provide for increasing responsibility as the girls grow older."[9]

The Technology Goddesses program has six goals:

1. To help girls acquire technology skills in the collaborative fashion they prefer, while helping to spark and sustain their interest through the years.
2. To sustain interest and involvement in technology from an early age.
3. To communicate the societal value of technology.
4. To imbue a sense of belonging and technical confidence.
5. To communicate the richness, variety, and fulfillment of IT careers.
6. To work with IT employers to share the unique needs of women in the workforce.[10]

Technology Goddesses offers a variety of engagement activities for young girls: one-day workshops such as "Computers in Everyday Life," "Web Design," and "Graphic Communications"; and field trips to museums and businesses in the area to talk about careers that utilize STEM skills.

But the marquee Technology Goddesses program is the one-week camps, held twice during the summer (where, Carmody said, "I sleep on a floor mattress"), that focus on the six goals of the Technology Goddesses program.

Each year Carmody "easily spends hundreds of hours" on the program. Jacobs Engineering, her employer, "has been tremendously supportive of Technology Goddesses. They love what I do and see the benefit of it."[11]

Since Technology Goddesses was launched 10 years ago, nearly 3,000 young girls have participated in the program. A tangible example of Technology Goddesses success is this: two young women from the inaugural 2002 class of the Technology Goddesses camp have earned STEM college degrees and have joined Carmody's IT staff at Jacobs Engineering.

I asked Carmody for her advice to IT executives interested in starting any kind of STEM program. She told me, "My best advice is—do anything. Lots of people stop themselves and say, 'What can one person do?' or 'I am too busy, I work full-time, travel internationally, and am raising a family.' I used to use those excuses myself. I was wrong. There is always more time in the day; making a difference can take many forms."[12]

Carmody also shared with me what the involvement with Technology Goddesses has meant to her. She told me, "Wow. This has been the most fulfilling, most stressful, and hardest thing I have ever done. It has honed my program management and organizational skills, has given me an opportunity to show that basic leadership approaches work outside the workplace, and has given me a tremendous impetus to stay on top of technology as it continues to change."[13]

You can follow Technology Goddesses on Twitter @techgoddesscamp and go to www.goddesscamp.org to review a wide array of informational materials

Carmody posts on the site—information that may inspire you to contact your local Girl Scout organization to start a Technology Goddesses program in your community.

Technology Goddesses brings a whole new meaning to Girl Scout "cookies"!

2002: nPower

Christopher Wearing was a senior managing partner at Accenture in 2002 and was very involved in the firm's philanthropic efforts. His prior experience with philanthropy, however, taught him a key lesson: money and technology donated to nonprofit groups often goes to waste. I asked Wearing, who is now the managing director of reengineering services at JPMorgan Chase, why this was the case. He told me, "Nonprofits often get access to in-kind donations of computer technology, networks, and software, but once they receive it they don't know what to do with it. I reasoned that nonprofits couldn't afford to hire a firm like Accenture, so I created nPower as an Accenture for nonprofits to help them with their tech challenges."[14]

Eleven years later, nPower, headquartered in Brooklyn, New York, has matched the skills of thousands of technology volunteers with nonprofit organizations, schools, and libraries in all 50 states.

Stephanie Cuskley, a former investment banker who joined nPower from JPMorganChase in 2007, is now its president, and a group of 13 prominent CIOs serve as the organization's advisory board. nPower offers four programs. I will focus on two of them, the Technology Service Corps and the Community Corps, because they have the most appeal and opportunity for involvement for IT professionals.

The Technology Service Corps is a highly selective, 22-week IT workforce development training program. It serves young adults (18–25) who have earned a general education degree, a high school diploma, or a certificate from the Cisco Networking Academy (see Chapter 13). The curriculum includes IT essentials such as hardware, software, web development, virtualization, customer service, and networking. Classes are taught by volunteer IT professionals.

A core component of the program is an eight-week paid internship. After program completion, 85 percent of Technology Service Corps participants are placed in jobs, often with corporations that underwrite nPower.

An nPower research survey found that nonprofit groups often identify technology as one of their most mission-critical needs, yet 40 percent said they lack sufficient technology and technology skills to service their constituents.[15]

Meeting that need is the purpose of the Community Corps. The program delivers critical IT and business support by matching the pro bono IT and business skills of volunteers to nonprofit organizations and schools that have requested assistance. Community Corps volunteers mostly participate in "short, simple, and impactful" projects that require 2–25 hours of volunteer time. That's important for busy IT executives, because it allows them to participate in a project, offer guidance and assistance, and manage the project through to completion in a relatively short period.

Since its inception, nPower has matched the skills of 1,800 IT executives with 48 foundations and nonprofit groups across the United States. It has definitely fulfilled Christopher Wearing's goal to bring the skills of big consulting firms to nonprofit and other deserving organizations.

Log on to www.npower.org for details on how you can volunteer your time to an nPower project.

2003: The Microsoft Imagine Cup

IT professionals enjoy competition. I experienced this firsthand in 1993 when Computerworld partnered with Microsoft and the Comdex, the big computer exposition, to create Windows World Open, an annual competition that challenged IT professionals to create custom applications for the Windows platform.

I saw it again several years ago when two Microsoft employees visited my office at *CIO* magazine to introduce me to the Microsoft Imagine Cup, an annual global technology competition for students who are 16 or older. The Imagine Cup was launched in 2003, and in the past decade it has accepted entries from 1,650,000 students hailing from 190 countries. (The United Nations recognizes 193 countries, so the Microsoft Imagine Cup has pretty much covered the world.)

The best way to describe the program is to review the 2013 competition. Every year Microsoft evolves the theme of the competition. The theme of the 2013 Microsoft Imagine Cup competition is "All dreams are now welcome." There are two rounds of competition. In the "local" (i.e., national) competition, one team is selected to represent that country in the Worldwide Finals. Microsoft pays all expenses to fly the winning national teams to the site of the global competition, which in 2013 is being held July 8–11 in St. Petersburg, Russia.

There are three broad areas of competition for the 2013 event: (1) "World Citizenship," which will recognize a software application, developed on a Microsoft platform, that exhibits the greatest potential to have a positive effect on humanity, (2) "Games," in which the winners will be honored for developing the most engaging and entertaining games built on Microsoft platforms (Windows 8, Windows Phone, Kinect for Windows Software Development kit, and xbox Indie Games), and (3) "Innovation," which rewards outstanding work in social media, e-commerce, and Internet search built with technologies like Windows Azure, Windows 8, and Windows Phone.

The 2013 Imagine Cup comes with the biggest cash prize allotment in the contest's history: $50,000 for first prize, $10,000 for second prize, and $5,000 for third prize.

Imagine this: If you know some 16-year olds who are gifted at creating innovative products with technology, encourage them to log on to www .imaginecup.com to register and join future Imagine Cup competitions.

2004: Engineering Is Elementary

When I founded Tech Corps in 1994, most of the organization's programs focused on high schools; less attention was given to sixth and seventh grades. In the mid-1990s, few even thought of teaching STEM subjects to children before sixth grade.

But the Museum of Science in Boston did, and in 2004 the museum launched the National Center for Technology Literacy to specifically focus on the T (technology) and the E (engineering) in STEM. Ioannis N. Miaoulis, the president and director of the Museum of Science, believed that "engineering is the great connector that uses math and science to solve real problems and often create new technologies that fuel innovation."[16] In 2004, the museum launched Engineering Is Elementary, an innovative curriculum program founded by Christine M. Cunningham, who remains the program's project director.

The Engineering Is Elementary curriculum program has five pillars: (1) children are naturally inclined to tinker and create, (2) engineering and technology literacy are essential for the twenty-first century workforce, (3) engineering in school holds the promise of improving math and science achievement by making all three subjects relevant to children, (4) children are capable of developing sophisticated skills and understanding in engineering at an early age, and (5) engineering fosters problem-solving skills.[17]

The 20-unit Engineering Is Elementary curriculum, which can be purchased by teachers and administrators, is not an independent curriculum that teachers have to learn before they can use it in their classrooms. Rather, the curriculum is simply paired with commonly taught science topics in the classroom.

I asked Cunningham to share with me comments from elementary school teachers who had used the Engineering Is Elementary curriculum program.

One teacher from the Chicago public school system, the nation's third largest, said, "For the past year I have been implementing [Engineering Is Elementary] as the first STEM-focused elementary school program within Chicago Public Schools. I have been beyond impressed with the curriculum, its reality in connection to the engineering and science worlds, and the excitement that it has brought to the diverse student body I teach in kindergarten through third grade."[18]

You read that last sentence right—kindergarten! Engineering Is Elementary is a successful example of how every school in America can introduce STEM concepts to students as early as kindergarten.

Engineering Is Elementary is also having a positive impact on educators. Cunningham quoted a teacher who wrote, "Wow! Engineering is so simple. I have been doing it all my life." And another said, "With Engineering Is Elementary, a child can be an engineer without knowing it. And to think [the child] could imagine to become an engineer as an adult and have a respected and lucrative career, well, that thought may not have ever occurred to [him or her] without exposure to Engineering Is Elementary. I teach inner city youths and would love to see an engineer emerge from my classroom one day. All because they have been exposed to the possibility of engineering at an early age."[19]

I could fill this book with equally glowing comments about the merits of the Museum of Science's Engineering Is Elementary program. But I think you get my point. This program works!

Log on to www.mos.org/eie to learn more about this innovative program and how you can implement it in the kindergarten at your local school.

2004: The Junior FIRST Lego League

I went to St. Louis, Missouri, for a speaking engagement. I like to arrive the day before a public speaking engagement for two reasons: (1) to avoid travel delays, and (2) to see the room I am slated to speak in the next day so I can envision myself onstage.

After arriving the day before in St. Louis and checking out the venue, I headed back to my hotel room to unpack and explore downtown St. Louis. As I walked through the hotel lobby, I noticed scores of young people, huddled in groups, sitting on couches and chairs, each group sporting a different T-shirt. I thought nothing of it as I approached the elevator and entered with a woman who was also wearing a colorful T-shirt, with JUNIOR FIRST LEGO LEAGUE across the front.

As the elevator door closed, I asked her why there were so many students in the lobby. "Oh, we are here for the annual FIRST Global Robotics Competition," she said. "The competition is being held across the street at the St. Louis Convention Center. You should stop by."

And that is exactly what I did. With flags from scores of countries hanging from the ceiling, the convention center looked like a special robotics session of the United Nations. Scattered across the floor were hundreds of teams setting up their booths and their special robotic machines for the competition that was to begin with a concert later in the day. It was exhilarating just walking through the aisles.

The organizer of the event was For Inspiration and Recognition of Science and Technology, or FIRST, which was founded in 1989 by Dean Kamen, the inventor of the Segway electronic transporter. The mission statement is uplifting: "Our mission is to inspire young people to be science and technology leaders, by engaging them in exciting mentor-based programs that build science, engineering, and technology skills, that inspire innovation, and that foster well-rounded life capabilities including self-confidence, communication, and leadership."[20]

A core component of FIRST is "gracious professionalism," which Woodie Flowers, a FIRST national advisor, describes as "a way of doing things that encourages high-quality work, emphasizes the value of others, and respects individuals and the community."[21]

FIRST started the global robotics competitions for high school students in 1992. But what captured my curiosity was a program that FIRST launched in 2004 called the Junior FIRST Lego League, developed for children ages six to nine, or roughly first through third grade.

The Junior FIRST Lego League is one of four programs offered by FIRST. The other three are the FIRST Robotics Challenge (which I had seen in St. Louis), the FIRST Technology Challenge, and the FIRST Lego League. The Lego Group, a global company focused on the development of creativity in children, is the founding sponsor of both the FIRST Lego League and the Junior FIRST Lego League.

When I returned to Boston, I arranged a telephone interview with Dana Chism, the program manager for the Junior FIRST Lego League. I asked her to tell me about the league. She explained, "First of all, unlike the other FIRST programs, Junior FIRST Lego League is *not* a competition, and it is *not* a science fair. Rather, it is a local-based program in partnership with the Lego education division that aims to teach the building blocks of engineering to six- to nine-year-olds."[22] Junior FIRST Lego League is mostly an after-school program.

Here's how the Junior FIRST Lego League program works. Every year FIRST and the Lego education division choose a theme for the program. In 2012 the theme was healthy eating. Teams, formed locally by parents or interested IT professionals (hint, hint!), must accomplish two tasks. Through a collaborative process, each team must draw a "show me poster" that illustrates the concept and research endeavor of the team. In essence, it is a drawing or model of what they want to accomplish.

The team must then build the illustrated model using Lego pieces, and it must include a simple motorized engine—quite a task for six- to nine-year-olds.

I also had an opportunity to speak with Trisha McDonnell, a vice president of Lego's education division, about the company's decision to become a founding partner of the FIRST Lego League and the Junior FIRST Lego League. McDonnell said, "It is all about bringing the passion of science and engineering to young people. By third grade it is critical to get children interested in STEM careers."

I asked her about the gender goals of the Junior FIRST Lego League, since my work with Tech Corps and other nonprofit groups had impressed upon me the importance of introducing STEM concepts at a young age to girls. "It is absolutely critical," McDonnell said, "to get *both* boys and girls involved to set the proper perception of what STEM careers are all about."[23]

Here's an important Junior FIRST Lego League fact that IT executives should consider: Although the league was launched in the United States, after just nine years, 40 percent of the new programs have been implemented in other countries. With all the increased competition that U.S. businesses now face from global firms, this underscores the importance of IT professionals getting involved in Junior FIRST Lego League.

I asked McDonnell about her favorite vision for Junior FIRST Lego League. She paused for a moment and then told me about a program in Taiwan where the team members had built a sushi refrigerator truck with Lego pieces—not with a motorized engine to power the truck, but with a motorized elevator on the back of the truck to lower the sushi to ground level for delivery.

I encourage readers to log on to www.juniorfirstlegoleague.org to get details on how your community can start a Junior FIRST Lego League program in your town.

2005: Raytheon's MathMovesU

The Raytheon Corporation deserves credit for its commitment to STEM topics. In my opinion, the company's efforts to build greater awareness and appreciation of mathematics among young people, particularly those in elementary and middle school, are among the best in the country.

Since 2005, Raytheon has invested $60 million in the MathMovesU initiative, which is the centerpiece of a multifaceted program that includes unique mathematics education projects, partnerships with leading STEM organizations, math scholarships for talented students, and volunteerism among employees. The company is also the underwriter of provocative research reports on the perception of mathematics among students, parents, and school administrators (referred to earlier in this book).

Raytheon approaches STEM education from many directions. The web site www.mathmovesu.com is the central repository of all the company's efforts. MathMovesU spreads the word about the fun of math through social media channels like Facebook (www.facebook.com/mathmovesu) and Twitter (@mathmovesu), where it hosts STEM discussions, shows educational videos, and has fun with items like Pi Day badges.

In partnership with the Smithsonian Institution in Washington, D.C., in the spring of 2012 the company opened the MathAlive! exhibit, which offers young students and adults hands-on experiences with mathematical concepts with the goal of giving them the opportunity to connect with math on a personal level. Because of the multiyear commitment that Raytheon has made to the Smithsonian Institution, it is estimated that 4 million people will experience the MathAlive exhibit. That's critically important to William H. Swanson, the chairman and CEO of Raytheon, who says in the company's MathMovesU brochure that "sometimes all it takes is a single moment, or spark, to inspire a future engineer to pursue a STEM career."[24]

Raytheon's MathMovesU initiative does an extraordinary job making a potentially boring subject like math exciting.

The "wow" factor was a prime reason that Raytheon partnered with Walt Disney Imagineering to create a unique program called the Sum of All Thrills at Innoventions at EPCOT, housed at the Walt Disney World Resort in Florida. Sum of All Thrills allows visitors of all ages to custom-design their own virtual thrill rides with unique mathematical tools. Then it gives them the opportunity to experience their creation in an immersive robotic simulator. That surely must make math come alive!

Each year Raytheon employees influence more than 100,000 middle school students, teachers, and parents through MathMovesU employee volunteer efforts. Raytheon's example of a broad-based commitment to improving STEM education in America should be an inspiration to every reader of this book. MathMovesU is one of the best programs in the country, if not in fact *the* best.

Can't get to Florida in the coming months? No problem! Just log on to www.mathmovesu.com. You can create a similar Sum of All Thrills experience on your home computer and learn more about how to incorporate the program into your community.

2005: IBM's Transition to Teaching

In the next decade, millions of teachers, many of them baby boomers, will retire. That exodus creates a huge opportunity for America to recruit a new generation of math and science teachers.

As noted earlier, a significant percentage, often more than 40 percent, of math and science teachers, particularly in middle school, teach those subjects out of field, which means that they do not have undergraduate or graduate degrees in science, technology, engineering, or math.

In 2005, the IBM Corporation took a hard look at this issue. With more than 100,000 employees in the United States, many of whom have worked at the company for 20 or more years and have postsecondary degrees in STEM disciplines, IBM began to think about how to engage those workers in a thoughtful discussion about their future plans—at the company or in a second career.

That's when Stanley Litow, the president of the IBM International Foundation and vice president of IBM Corporate Community Relations, came up with the idea for Transition to Teaching. At the launch of the Transition to Teaching pilot program in 2005, Litow explained the program's goal: "Many of our experienced employees have math and science backgrounds and have made it clear that when they are ready to leave IBM they aren't ready to stop contributing. They want to continue working in positions that offer them the opportunity to give back to society in an extremely meaningful way. Transferring their skills from IBM to the classroom is a natural for many— especially in the areas of math and science."[25]

Sounds like a great program, doesn't it? It is, and it remains one of my favorite STEM initiatives in the country. If it were adopted by hundreds of other companies, it has the potential to recruit tens of thousands of new math and science teachers—teachers who actually have degrees in science and math—to replace the retiring baby boomer teachers.

Maura Banta, the director of citizenship initiatives in education for IBM, testified about the merits of the Transition to Teaching program to a House subcommittee in 2012. She shared the components of the program's success:

> IBM launched Transition to Teaching to address the K–12 STEM pipeline issues by facilitating retiring IBMers moving into science and math education as a way of helping to encourage young people to enter STEM careers. More than 120 of our most experienced employees have participated in the Transition to Teaching program. Each person who was chosen for the program is a math or science professional with at least one degree in a STEM field. . . . Most program participants have engineering backgrounds. . . . As part of Transition to Teaching, they participate in a wide range of teacher certification programs (which can take up to four years to complete), and the specific licensing requirements required by the state in which they live.
>
> Since teachers must have strong, in-depth backgrounds in their subject areas, we focus on IBMers who have a bachelor's degree or

higher in a math or science discipline. Because we believe IBMers need to learn the craft and skill of teaching, classroom management, and instructional practice to be effective educators, we reimburse their tuition costs for their education preparation. IBM provides stipends up to $15,000 so those who are transitioning to teaching can take leaves of absence—while maintaining their IBM benefits—to do student or practice teaching.

In our Transition to Teaching experience, at least three challenges must be addressed in order to attract math and science professionals to education and prepare them to become exemplary teachers. We encourage: (1) the development of standards for the pedagogic and instructional skills and knowledge required and focus only on those education courses that are necessary for teacher certification, (2) assurances that teaching candidates are placed in supportive practice environments under qualified instructors, and (3) systems that will provide new teachers with mentoring and peer support during their first two years to ensure that they are able to provide the highest-quality education to their students.

If an additional 25 large companies established programs similar to Transition to Teaching, their combined efforts could provide a substantial number of new math and science teachers.[26]

Banta's testimony before Congress gives you the magic ingredients to start a Transition to Teaching program at your company.

Litow's vision in creating Transition to Teaching was prescient. According to the National Science Foundation's 2007 *Elementary and Secondary Mathematics and Science Education* report, 19 percent of all public middle and high school mathematics teachers, and 22 percent of all public middle school and high school science teachers, entered the profession from alternative routes like the workplace.[27]

2006: The Khan Academy

Through what medium do you think children would prefer to learn math and science: 300-page textbooks or short videos on YouTube?

Salman Khan figured out the answer to that question in 2006 when he launched the Khan Academy, an online repository of more than 3,400 video tutorials. These videos, housed on both the Khan Academy web site and YouTube, feature simple, no-frills instruction in mathematics, history, medicine, physics, biology, chemistry, computer science, and other topic areas.

You may already be familiar with the Khan Academy. Salman Khan has been interviewed widely about his innovative teaching model that offers free world-class education. The Khan Academy model is very democratic. It is open to students, teachers, homeschoolers, principals, and adults who want to learn. All materials and resources are free of charge.

Students can make use of the extensive Khan Academy library of micro-instructional videos. They can participate in interactive challenges and get instant feedback on how well they are doing.

Let's say your 14-year-old daughter is having problems with algebra in school. Logging on to the Khan Academy, she can search for algebra tutorials, click on the video link, and become immediately engaged in a simple video format, often narrated by Khan. As she proceeds to more challenging videos, perhaps on the same algebra topic, she earns badges and points as rewards for her progress.

To prevent the system from being too predictable for the students, all content is randomly generated, like ads served up on web sites, so most students never run out of material to interact with.

For many, including myself, the Khan Academy model is about much more than an innovative online learning experience. It begins to question the very model of learning and education. Let me explain.

About a year ago I attended MIT's CIO Symposium, where one of the keynote speakers was Anant Agarwal, the former director of MIT's computer science and artificial intelligence laboratory. On this day, however, he was talking about his new role as the first president of edX, a joint online learning experiment MIT was conducting with Harvard University.

The edX business plan set out to attract about 2,000 students in an elementary physics class taught online by 10 professors from the two institutions. Soon after the launch, however, the course enrollment skyrocketed to more than 100,000. Agarwal asked his MIT audience, "How am I going to answer all the questions those students may have taking the course?"[28]

He didn't have to worry about that, however, because of the power of social media. That is, the students themselves answered the other students' questions rather than waiting for a professor to respond.

Agarwal surprised the audience when he asked, "What does it mean to have a college education?" He continued, "Right now the model is predicated on the notion that after four years at any given institution, if you complete the courses that we the faculty say are important, you earn a piece of parchment with your name on it."[29]

The inference, though not actually spoken on the stage that day, was the model playing out in this edX experiment. It had the power to entirely change the world of higher education, an out-of-touch world where tuitions and fees always go up, year after year.

And at the other end of the education spectrum, Salman Khan's model is evolving in a similar fashion. Khan and Agarwal, through utilizing the web, are creating incredibly powerful new business models for education.

An op-ed in the *Wall Street Journal* noted that the education models of edX and the Khan Academy are pushing education to "a crossroads not seen since the introduction of the printing press."[30]

If you have kids, or if you just want to brush up on your math and science skills, log on to www.khanacademy.org and take a few classes. If you want to see the future of college education, go to the MIT and Harvard site www.edx .org.

2006: Cognizant's Maker Faire

My five years of STEM research has introduced me to hundreds of efforts by companies and nonprofit organizations. As I was finalizing my list for the book, I had lunch with Mark Greenlaw, the vice president of Cognizant Inc., a provider of IT, business process, and consulting services located in Teaneck, New Jersey, to review my list. Mark listened politely, and when I had finished my comments on the book, he asked, "Have you heard of Maker Faire?"

"Maker what?" was my response. But two hours after lunch had ended, I knew what Maker Faire was. And I was very impressed.

Let's start by describing its mission: "The Maker Faire Initiative is to create more opportunities for young people to make, and, by making, build confidence, foster creativity, and spark interest in science, technology, engineering, math, the arts, and learning as a whole."[31]

Maker Faire was launched in San Mateo, California, in 2006 by Dale Dougherty, now the general manager of the Maker Media, formerly a division of O'Reilly Media. The first fair consisted of six workshops, six exposition tents, and 100 Maker exhibitions spread throughout a five-acre tent. The 2012 World Maker Faire, held at the New York Hall of Science on the grounds of the 1964 New York World's Fair, featured 500 Maker exhibits, including 3D printers, green tech, home energy and robotics, scores of speeches by notable individuals in the technology community (such as Chris Anderson, the editor-in-chief of *Wired* magazine), and hundreds of workshops. Well over a hundred thousand people walked through the fair.

Dougherty's goal in starting Maker Faire was "to look at things a little differently and spark the next generation of scientists, engineers and Makers."[32]

Getting interested? Are you involved with a youth group? Maybe your son or daughter is good at making things. Go to www.makerfaire.com and click on the "how to participate" icon in the navigation bar. The World Maker

Faires are usually held in the fall and have a deadline of late August. Follow the instructions on the entry form. (Hint: Maker Faire first-round judges are particularly impressed with projects that are "interactive" and "highlight the process of making things.")[33]

Cognizant donated $100,000 to the 2012 World Maker Faire in New York City in the form of college scholarships ($5,000 to each of 20 winners at the fair) to students interested in enrolling in STEM programs at the university level. The company was also the primary sponsor of the Young Makers Pavilion at the World Maker Fair, an exhibit designed to showcase the work of middle school participants.

So satisfied is Cognizant with Maker Faire that it is taking it to all parts of the country through a traveling road show called Making the Future, an after-school and summer program aimed at building creative, innovative, and collaborative skills in young Americans.

Greenlaw explained his company's broad commitment to Making the Future this way: "The program seeks to encourage exploration, invention, and the joy of learning" in young people.[34]

Making the Future has other components to it, including financial support and more than 100,000 hours of Cognizant employee volunteer support for nonprofit organizations like Citizen Schools and the Boston Museum of Science's Engineering Is Elementary program. Greenlaw and Francisco D'Souza, the firm's CEO, are active participants in the Business Roundtable and Change the Equation, two key CEO-led initiatives that cultivate widespread STEM literacy efforts across America.

The Maker Faire concept is gathering the attention of lots of very big companies, like Radio Shack, Red Bull, the Walt Disney Corporation, Time Warner Cable, ASUS, Autodesk, Crayola, Delta Faucet, Schick, SketchUp, and General Motors.

2007: The National Math and Science Initiative

The National Math and Science Initiative (NMSI), a nonprofit organization launched in 2007, was created as a direct response to the 2005 National Academy of Sciences landmark *Rising above the Gathering Storm* report (reviewed in Chapter 11).

The NMSI is funded largely from contributions from the ExxonMobil Corporation (which frequently airs TV commercials featuring golfer Phil Mickelson touting the benefits of science and math education), the Bill and Melinda Gates Foundation, the Michael and Susan Dell Foundation, and the Texas Instruments Foundation. Some look at the NMSI as the "bank" of the STEM movement in the United States.

But it is much more than that. The NMSI aims to make a difference through an approach called "scale philanthropy," an approach to giving that identifies math and science education programs with proven results. The NMSI looks for the best ideas and the best programs and works closely with the founders to spread their ideas nationally.

Three key programs the NMSI supports are the UTeach Institute, an innovative math and science teacher recruitment program that began in 1997 at the University of Texas at Austin; the AP Training and Incentive Program, which aims to dramatically improve college readiness by increasing the number of high school students who take and succeed in advanced placement math and science tests; and Laying the Foundation, which is dedicated to improving the quality of teacher training.

The group's driving mission is that "we don't reinvent wheels, we find the best ones—and roll."[35]

There are several ways you can become involved with the NMSI. First, if you know of an outstanding local science, technology, engineering, or math program in your community—with proven results and the potential to spread—have it apply for a grant from NMSI. Second, from its six years of work with STEM initiatives, the NMSI has come up with a list of ways Americans can support math and science education. Here are the suggestions that I found most interesting:

1. Think and speak positively about math and science. . . . Never say to a child, "I wasn't good in math, either." Rather, say, "Learning math is critical for everyone today. I sure wish I had studied it more."
2. Pay close attention to math and science teaching in your child's school. Look in particular to see if assignments and projects are creative and tied to real-life situations.
3. Offer to mentor students in science and math.
4. Volunteer to help organize a science fair if your middle school does not have one.[36]

The web site for the NMSI is www.nms.org.

2008: AT&T Aspire

I was sitting in a hotel conference room in Chicago at a meeting where AT&T CEO Randall Stephenson was speaking to a group of CIOs. During his remarks, Stephenson referred to the AT&T Aspire program, and during the question-and-answer session that followed his remarks, I asked him to share more details about Aspire with the group. An amazing story followed.

Stephenson became the chairman, CEO, and president of AT&T in June 2007. With outsourcing being then, as it is now, a hot labor topic across the nation, he had an idea: he wanted to create 5,000 new AT&T jobs in America. Specifically, he wanted to hire 5,000 technical workers to help the company install U-Verse, the television, telephone, and Internet service that the company had commercially launched the year before.

That's when reality set in. As Stephenson answered my question, he said, "It is easier to be accepted into an Ivy League university than it is for AT&T to find and hire a worker with the basic tech skills needed to install our U-Verse system." (His rationale was that the overall acceptance rate among the eight Ivy League schools was 13 percent at the time, whereas AT&T was finding only one qualified candidate out of 10 or 11 applicants, making its "acceptance" rate around 9 percent.)

So in the second half of 2007, Stephenson worked with his team, and on April 17, 2008, AT&T introduced Aspire, a $100 million philanthropic effort to improve workforce readiness in the United States. At the program's launch, Stephenson said that "1.2 million students drop out of high school every year. This has implications for individuals and our nation's global economic leadership. AT&T Aspire is about supporting the great work already underway to help our kids succeed in school, and helping students see the connection between education and their best future."[37]

The three pillars of the initial AT&T Aspire program were the following: (1) the distribution of grants to schools and nonprofit groups focused on helping students to graduate from high school, (2) a massive student job-shadowing initiative, involving 13,000 AT&T employees who donated 286,000 hours and affected 105,000 students, and (3) financial underwriting support for national research on technology and high school dropout programs.[38]

AT&T upped the stakes and in 2012 announced an additional $250 million financial commitment to Aspire, spread over five years, to focus on using technology to connect students in new ways, such as interactive gaming, Web-based content, and social media.

I interviewed several AT&T employees and asked them about their volunteer experience with Aspire. Marko Popovik, the strategic business process modeling manager for AT&T in Chicago, organized a high-tech day for 100 high school students in the area. I asked him what was his biggest impression from his involvement with Aspire, and he told me, "The impact on the students. After reviewing the survey results from the event's participants, almost all of the students said this inspired them to move forward with education careers and to stay in school to achieve their goals. This was a tremendous achievement, knowing that most students in these schools had the largest dropout rates in the city."[39]

Ebony Williams, the senior technical leader for IT corporate systems, volunteered her time to Aspire "because I wanted to do my part to help reduce the escalating high school dropout rate. There are so many students who desperately need adequate mentors and role models. The students I met, the experiences they shared with me, and the hope they expressed to me for their futures was not only inspirational, but it also left me with a desire to serve in a greater capacity."[40]

In October 2012, AT&T Aspire significantly expanded its job-shadowing, mentoring, and internship components with the launch of the Aspire Mentoring Academy, which aims to challenge AT&T employees to donate 1 million hours to help students graduate from high school and succeed in college or the workplace. The Aspire Mentoring Academy supports three initiatives:

1. Job mentoring, where employees share life experiences and career advice through project-based activities during the workday.
2. Skills mentoring, where employees share academic and life skills with students
3. E-mentoring, which will utilize technology to enable employees to share their math skills and underscore their relevance in the real world through online mentoring.[41]

Five years after AT&T Aspire had been launched, I asked Stephenson for an update on his perception of STEM challenges in America and specifically what Aspire means to him.

In addressing the broad STEM challenges, Stephenson told me the following:

At AT&T, our success depends on a skilled, well-educated workforce. But it's not always easy to find highly qualified candidates for the kinds of high-tech jobs we create. And we're not alone. Across many industries, there are too many high-paying jobs that are going unfilled—while unemployment remains stubbornly high. Unfortunately, the future won't take care of itself, because a majority of the job openings will come in areas with predicted talent shortages, especially in the STEM fields. We need to close this talent gap and I'm confident we can—if we invest our time and our dollars in proven programs that help high school students who may be at the risk of dropping out.

Asked for his thoughts five years after founding Aspire, he said this:

AT&T Aspire is all about working hard to make classroom learning come to life for high school students, and helping them to make a connection between what they learn in school and a good career. We're doing this by leveraging the innovative technologies that are at the heart of our company, and by harnessing the collective power of our employees who participate in the Aspire Mentoring Academy. Our goal is very straightforward—we want to inspire and equip these young people with the tools to graduate and the skills needed to build good careers. If we're successful, we'll have a real impact on the lives of students across the country, and the payoffs for our economy and our society will be huge.[42]

The story of AT&T Aspire an interesting one for IT executives because it shows how one individual—one person with a plan, a passion, and a vision—can rally people to a cause and get things done.

Granted, the story here is about that one person being the CEO of AT&T. He's in a job that has access to lots of resources. But that's not the main point; how Stephenson responded to that challenge is. One person *can* make a big difference. I hope Stephenson's example of leadership inspires you.

2008: AMD's Changing the Game

A 2009 high school survey of student engagement reported that "two-thirds of American high school students report being bored in class every day."[43]

But AMD—a global semiconductor design company with a strong product position in graphics technology used in a wide array of digital devices, ranging from game consoles to personal computers to servers—saw an opportunity in that horrid high school statistic.

In 2008, the AMD Foundation, the philanthropic arm of AMD, launched Changing the Game as a funding initiative with the purpose of taking gaming technology beyond entertainment. According to the AMD Foundation, the goal of Changing the Game is to teach and challenge young people to create their own video games with social content, and in doing so learn critical STEM skills as well as problem solving, critical thinking, and collaboration.[44]

Earlier I addressed a major challenge America faces in getting more young people interested in STEM: they find the subjects boring, difficult, and not particularly relevant to their lives.

Changing the Game addresses that challenge head-on by funding programs like Gamezone on Whyville, the U.S. National STEM Video Game Challenge, a nationwide competition that invites game makers of all ages to show their passion for both playing and making video games, and by supporting Change the Equation, a nonprofit group that I will describe later in this chapter.

Allyson Peerman, the vice president of global public affairs at AMD and the president of the AMD Foundation, shared the following with me: "The success of Changing the Game is due to one factor: the programs the Foundation supports get young adults excited again about education and motivate them to find their passion about technology."[45]

In 2012, the Computerworld Honors Program honored AMD's Changing the Game initiative with a prestigious Laureate award "for the Foundation's use of technology to promote and advance the science, technology, engineering and math skills of youth around the world."[46] And Peerman has even bigger plans. "In the first four years the program has influenced the lives of 75,000 young people around the world," she told me. "By 2020 our goal is to increase that to 1,000,000 children."[47]

When I speak with executives like Peerman, I often ask, "In addition to dollars, why are employees at your company getting involved?" I sent her a list of questions that she shared with employees who had volunteered their time to the Changing the Game program.

She wrote that Diane Stapley, a senior product marketing manager at AMD said, "I've worked with several professional game development companies in my life, and these sixth and seventh graders' games are on par with some of the best I've ever seen. I can only imagine how fluent these children will be when they enter the workforce in 10 years!"[48]

Another comment Peerman forwarded was from Ward Tisdale, the director of global community affairs at AMD, who said, "It's been a great pleasure to see Changing the Game live up to its billing in places like Beijing, Austin, and Bellevue. We've only begun to see the potential of youth game development as a way to improve STEM learning and acquire 21st-century skills."[49]

A third employee, Anne Feritta, a senior manager at the company, said, "I am amazed at the level of technical knowledge the students possess. It isn't often you see kids get excited about math. Watching the kids was an eye-opener. They were collaborating, brainstorming, and researching how to teach students about fractions. Not only were they helping classmates learn, they were also learning more about math in the process."[50]

AMD Foundation's Changing the Game initiative is successful because it makes it fun to learn about STEM, and it is giving tens of thousands of young people the opportunity to bring their "A game" to the game of life.

2009: Microsoft's TEALS

If there were more Kevin Wangs in the business world, the United States would not have a STEM crisis. One of the big issues with most middle school and high school science and math teachers, we have noted, is that they are out of field: they do not have an undergraduate degree in math, engineering, or science.

Wang is the polar opposite of an out-of-field science and math teacher. He graduated from the University of California at Berkeley with a degree in electrical engineering and computer science. One would suppose that Wang would stay in the San Francisco–Silicon Valley area to strike it rich.

Wrong! He opted instead to attend Harvard University, where he earned a master's degree in education. After graduating from Harvard, he headed back to the Bay Area to build a curriculum and teach computer science to middle and high school students, a job he held for three years. During those years, Wang was appalled by the fact that only 19,390 out of 14 million students in the United States opted to take the computer science advanced placement test in junior or senior year of high school.

Wang was perplexed. With some of the fastest-growing and highest-paying jobs in the marketplace requiring degrees in computer science or engineering, why were so few American students opting to take computer science in school?

After trying for three years to change the perception of computer science education within the walls of academia, Wang set out on a new course. He was recruited for, and accepted, a position at the Microsoft Corporation in Redmond, Washington. After several years at Microsoft, however, Wang realized he missed teaching, so he began to teach a morning computer science class at University Prep, a Seattle-based middle school.

And then Wang had his epiphany moment: Maybe he could do both!

His seminal observation focused on the work habits of software developers and tech workers. In a local newspaper interview, Wang explained, "Very few people come in to work before 10 a.m." and "With all the schools around that need computer science teachers, I realized there's a huge opportunity." Although Microsoft employees often work late into the evening, most have free time from 8 to 9 a.m.[51]

It was the proverbial win-win situation. In 2009, after a local high school that was in danger of losing its computer science program contacted Wang to volunteer to teach its computer science courses, he founded Technology Education and Literacy in Schools, or TEALS, and began to recruit volunteers to join him in teaching three computer science courses: Introduction to Computer Science, Web Design, and Advanced Placement (AP) Computer Science. Intro and Web Design are each one-semester courses, and AP Computer Science is a full year.

The key to the success of the TEALS program is that the classes are offered exclusively during the first period of the school day, allowing the talented Microsoft volunteers to participate. And talented they are. All Microsoft

TEALS volunteers have undergraduate engineering degrees, many have graduate degrees in STEM-related fields, and two TEALS volunteers even have PhDs.

The commitment that TEALS volunteers make is impressive. Every volunteer teaches his or her one-hour class before arriving at work—four days a week!

Rubaiyat Khan, an MIT-educated Microsoft program manager who contributes time to a TEALS computer science course at Lake Washington High School in Kirkland, Washington, explained her reasons for volunteering. "I am passionate about teaching. I love seeing the lightbulbs go on in students' eyes when they learn something completely new and then use that knowledge in life. It's amazing when you see a kid get it." Khan enjoys her dual role as teacher and role model to young women in the class. "One thing I like is that girls see us teaching. We need more role models of people—both men and women—with successful computer industry careers to teach in schools so students see that as a viable option for themselves."[52]

Corinne Pascale, who works on the Microsoft Office 365 team, said she "likes being able to relate the value of computer science education to the careers many of my students envision for themselves, such as forest service fields." The TEALS program gives back to volunteers, she noted, particularly helping her to hone her communication skills. Commenting on her prior reliance on geek-speak, Pascale added, "I was losing my ability to talk, like with my parents, in a nontechnical way. If I can explain to a 16-year-old what a recursive loop is, that helps."[53]

In the pilot school year of 2009–2010, TEALS was operational in one Seattle school. Only 12 students—none taking AP Computer Science—participated, and there was one TEALS teacher, Kevin Wang. By the 2012–2013 school year, the program's growth was remarkable: 37 schools in the Seattle area were participating, 2,000 students were enrolled in TEALS courses taught by 120 Microsoft employees, and 300 of the students were taking AP Computer Science. And even though TEALS is now Wang's full-time job at Microsoft, the company is recruiting other firms to join the cause. In a 2012 *New York Times* article about TEALS, Alyssa Caulley, a Google software engineer who works with a Microsoft partner to teach a computer science class, said about TEALS, "I think education and bringing more people into the field is something all technology companies can agree on."[54]

The commitment of TEALS volunteers continues to impress Wang, who said that the typical TEALS volunteer "works hard over the summer to train for the school year and then puts in just an astonishing number of hours of preparation—300-plus hours each year—including doing the thing that computer science people hate the most, getting up early."[55]

Several TEALS volunteers underscored Wang's comment about commitment. Colin Miller, a principal program manager at Microsoft, said that TEALS "is a huge time commitment to do right." Moreover, in terms of the challenge of improving how computer science is currently taught, Miller confessed that it "was more complex than I had thought originally. But like almost every engineer that I know, I have thought about teaching as an encore career, and TEALS seemed like a great way to try it out."[56]

Michael Hawker, a Microsoft software development engineer, said his experience with TEALS "really puts what you do in perspective. It's rewarding to stand in a classroom and have students look to you for advice and knowledge and know you've had an impact in their lives. Running a class of 30 students can call for a large commitment of time, but even if you have just a few hours to spare, TEALS is a great way to make an impact."[57]

Cecil Sidwell, a program manager for Windows Update, volunteered time to the TEALS program because "I have this desire to pass on the knowledge I have to the next generation." Sidwell also participates because "the world is changing rapidly, and our public education needs to change just as quickly. If we don't, the rest of the world will be leaps and bounds ahead of the United States."[58]

That hard work is delivering results. As most people know, advanced placement courses are rigorous high school programs designed to replicate the difficulty of college-level work. Each year the College Entrance Examination Board administers advanced placement tests that are scored from 1 to 5, with 5 being the best. The national average for students taking the AP Computer Science test is 3.14.[59] For students who are taught computer science in the TEALS program, the average score is 4.15—32 percent higher than the national average!

TEALS is an outstanding example of how the vision of one IT worker can positively influence so many lives. The TEALS program, though currently concentrated in the Seattle area, has expanded into Utah, California, and North Dakota and is planning to spread the program to the East Coast.

Asked about the reason for the program's success, Wang told the *New York Times*, "Kids see themselves in their shoes [the volunteers']; after all, the chances of going to college and majoring in computer science are exponentially better than getting into the National Football League."[60]

Isaac Wilson, a Microsoft software development engineer, said, "Even though TEALS is a huge commitment, the program is in an amazing position because there's huge demand for its classes and there is a huge potential supply of teachers. The key limiter to growth right now is the ability to spread TEALS while maintaining the quality of instruction. I am completely confident TEALS, and Kevin Wang's business model, can spread to the entire U.S. and make computer science a ubiquitous high school experience."[61]

Log on to www.tealsk12.org to start a program in your community.

2009: The Salesforce.com Foundation

Marc Benioff is a wildly successful technology business executive and one of the industry's most philanthropic visionaries. Moreover, Benioff is a prolific writer; he has coauthored four books.

My favorite Benioff book is *Compassionate Capitalism: How Corporations Can Make Doing Good an Integral Part of Doing Well* (2004), in which he and coauthor Karen Southwick explain the underpinnings of the company Salesforce.com's three-pronged approach to corporate philanthropy: "1–1–1."

Salesforce.com's philanthropic 1–1–1 model is built on these pillars: (1) 1 percent of the company's founding stock has been set aside to fund the Salesforce.com Foundation, (2) every Salesforce.com employee is encouraged to spend one (paid) week of every year volunteering at a nonprofit agency of his or her choice, (3) and every year, Salesforce.com donates 1 percent of its previous year's revenue in the form of Salesforce.com software.

Benioff's vision in setting up the Salesforce.com Foundation is simple and easy for employees and customers to understand, and it is working. Salesforce.com employees donated 75,000 hours in 2011 to 900 organizations.

Many companies, particularly software firms, donate software to nonprofit groups that cannot afford it. But as Christopher Wearing, the founder of nPower, explained earlier in this chapter, that approach often doesn't work because the receiving organization does not have the resources to install and maintain the donated software package.

Although the Salesforce.com Foundation employees can donate their time to any deserving group, many link their one-week volunteer effort to a local nonprofit group that has just received a donation of Salesforce.com software and needs tech expertise to get the program installed, operating, and maintained.

The Salesforce.com Foundation model can be replicated by any company, large or small. Donate 1 percent of your company's revenues to a philanthropic cause, and allow your IT workers to donate one week a year, with pay, to a volunteer cause.

Corporate philanthropy is good for the community, and it's good for business. Benioff's plan is brilliant because of its simplicity. Try it at your company.

2009: DIGITS

The leaders of the technology community in Massachusetts were shocked in the fall of 2009 to learn that only 22 percent of students taking the SAT in Massachusetts that year expressed interest in pursuing a college major in STEM, compared to 28 percent of SAT takers nationally.

Concerned that the state's education system would not be able to supply enough future STEM workers, six statewide science and technology associations met in Sturbridge, Massachusetts, at the First Massachusetts STEM Summit. They then launched DIGITS, an innovative program that pairs working STEM professionals with sixth-grade classroom students and teachers to increase student interest in STEM careers.

There are three main volunteer components of DIGITS: (1) hands-on science and math activities in the classroom, (2) the sharing of personal career stories and an explanation of what the careers are and how they use math and science, and (3) explanations of the attributes and benefits of working in a STEM career.

The overall goal is to inspire sixth graders that STEM careers are "interesting, engaging, fun, and socially 'cool.'"[62]

One STEM ambassador, as an IT executive participating in the program is called, was challenged with having to explain to sixth graders how big data storage systems work. He talked about an iPod, a music device most sixth graders are familiar with, and the fact that his company's data storage system could hold 250 million songs that would take 1,900 years to listen to![63]

Another STEM ambassador from a biotech company turned the sixth-grade classroom she visited into a mini-laboratory. She brought in microscopes and slides that showed what normal cells and cancer cells look like and then led a discussion about the causes of cancer and how scientists continue to fight the disease.[64]

Although DIGITS is currently a Massachusetts-based initiative, the idea will work in any community or state. Its focus on sixth graders is important because that is an age at which young people become interested in STEM careers.

DIGITS has successfully reached more than 36,000 students in 129 schools served by 400 volunteers from 90 corporations in just the first two years of the program.

Teachers and school administrators, who often are reticent to bring in outside experts to teach in their schools, support DIGITS, and the kids like it, too. One sixth grader thanked a STEM ambassador with this note: "Thank you for teaching us about science. I honestly don't know what I'm going to be, but I know it's going to involve science, math, physics, electrical engineering, mechanical engineering—or all of the above!"[65]

That's some influence. Log on to www.digits.us.com for more details about this exceptional STEM volunteer program.

2009: Change the Equation

Change the Equation is a nonprofit group focused on improving STEM education in America. This CEO-led group, with more than 100 corporate members from our nation's largest and most respected firms, commits more than $500 million each year to STEM education initiatives throughout America.

I call Change the Equation the "Smithsonian Institution of STEM data and statistics," because so many times, while conducting research for this book, I found myself landing on a Change the Equation page that clearly and visually presented the data I was looking for. All too often, STEM data are boring, dense, hard to decipher, and hard to find in one place.

Change the Equation takes a big-picture approach to STEM philanthropic efforts through a rather unique program called Design Principles for Effective STEM Philanthropy, which endeavors "to define a framework for corporate engagement that has the best chance of addressing the profound STEM challenge we face as a nation."[66]

There are six design principles:

1. Identification and targeting of a compelling and well-defined need.
2. Rigorous evaluation to continuously measure and inform progress.
3. Ensuring that work is sustainable.
4. Demonstration of replicability and scalability.
5. Creation of high-impact partnerships.
6. Ensuring program capacity to achieve goals.[67]

For readers who are interested in starting their own nonprofit STEM efforts, these six principles are your STEM nonprofit business plan.

The Change the Equation web site, www.changetheequation.org, contains a wealth of information about all things STEM. The STEMWorks Database is a guidepost for companies, or wealthy individuals, looking to invest in STEM programs that meet the rigorous provisions of Change the Equation's design principles.

Interested in how your state's STEM efforts are doing? Just click on the Vital Signs link for a comprehensive snapshot of the demand for, and supply of, STEM skills in your state and the state's STEM expectations of students and teachers. My favorite part of the Change the Equation site is STEMistics, an incredibly large library of recent STEM articles and statistics on issues like STEM in the economy, minorities in STEM, government policy, and women in STEM.

I have been involved with education technology nonprofit organizations for 20 years. I have seen the following story play out all too often: a company or an individual gets a good idea and funds it, but eventually the effort fails because the idea was not critically vetted at the initial stages of the project. Many companies and individuals want to do good, particularly in STEM-related initiatives; that's not news. But Change the Equation's core mission of defining the components of proven STEM initiatives—efforts that deserve continued investment and have a great opportunity to succeed—*is* news.

Notes

1. Anne Roder and Mark Elliott, *A Promising Start: Year Up's Initial Impacts on Low-Income Young Adults' Careers*, Economic Mobility Corporation, April 2011, www.yearup.org/pdf/emc_study.pdf.
2. Year Up, "Program and Results," n.d., www.yearup.org/about/main.php?page=program.
3. Gerald Chertavian, *A Year Up: How a Pioneering Program Teaches Young Adults Real Skills for Real Jobs with Real Success* (New York: Viking, 2012), 5.
4. Year Up, "Students," n.d., www.yearup.org/students_alumni/main.php?page=students&sub_section=national.
5. Ibid.
6. Juniper Networks, "Juniper Networks Foundation Fund Committee Overview 2012," e-mail to author, August 2012.
7. Stacey Clark Ohara, e-mail to author, August 23, 2012.
8. Technology Goddesses, www.technology-goddesses.org.
9. Cora Carmody, e-mail to author, October 8, 2012.
10. Technology Goddesses, "The Mission of Technology Goddesses Is," www.technology-goddesses.org/Technology_Youth.pdf.
11. Carmody e-mail.
12. Ibid.
13. Ibid.
14. Christopher Wearing, interview with author, New York, September 19, 2012.
15. nPower, "The Community Corps," www.npower.org.
16. Ioannis Miaoulis, "Museums Key to STEM Success," *U.S. News & World Report*, December 7, 2011.
17. Christine M. Cunningham and Cathy P. Lachapelle, "Engineering Is Elementary" (Boston) Museum of Science, 2010, www.mos.org/eie.
18. Christine M. Cunningham, e-mail to author, August 29, 2012.
19. Ibid.
20. FIRST, "Vision and Mission," www.usfirst.org/aboutus/vision.
21. FIRST, "About Us," www.usfirst.org.
22. Dana Chism, interview with author, Manchester, NH, August 13, 2012.
23. Trisha McDonnell, interview with author, Pittsburgh, KS, August 28, 2012.
24. William H. Swanson, "MathMovesU," Raytheon Corporation, 2012, www.mathmovesu.com.
25. IBM Corporation, "IBM Launches Transition to Teaching Program," press release, September 16, 2005.
26. Maura Banta, testimony to the House Subcommittee on Early Childhood, Elementary, and Secondary Education, July 24, 2012.
27. National Science Foundation, *Elementary and Secondary Mathematics and Science Education*, 2007, www.nsf.gov/statistics/seind12/pdf/c01.pdf.
28. Anant Agarwal, presentation at MIT's CIO Symposium, Cambridge, MA, May 22, 2012.
29. Ibid.
30. L. Rafael Reif, "What Campuses Can Learn from Online Teaching," *Wall Street Journal*, October 3, 2012.
31. Maker Initiative, "Every Child Is a Maker," www.makered.org.
32. Ibid.
33. Ibid.
34. T. J. McCue, "Making the Future with STEM Scholarships," *Forbes*, September 25, 2012, www.forbes.com.
35. National Math and Science Initiative, "Our Approach," http://staging.nationalmath.andscience.org/our-approach.

36. National Math and Science Initiative, "Other Ways to Help," http://staging.national
    .mathandscience.org/solutions/how-you-can-help/other-ways-help.
37. AT&T Corporation, "AT&T Launches $100 Million Philanthropic Education Program,
    Job Shadowing for 100,000 Students, Research and Community Engagement Support to
    Address High School Dropout Crisis" press release, April 17, 2008, www.att.com/gen/press-
    room?pid=4800&cdvn=news&newsarticleid=25507.
38. Ibid.
39. Marko Popovik, e-mail to author, October 17, 2012.
40. Ebony William, e-mail to author, October 17, 2012.
41. Janice Evans-Page, "One Million Hours, Unlimited Potential," *AT&T Consumer Blog*,
    October 1, 2012, http://blogs.att.net/consumerblog/story/a7784367.
42. Randall Stephenson, e-mail to author, November 2, 2012.
43. Ethan Yazzie-Mintz, *Charting the Path from Engagement to Achievement: A Report on the 2009
    High School Survey of Student Engagement*, Indiana University, 2010, http://ceep.indiana.edu/
    hssse/images/HSSSE_2010_Report.pdf.
44. AMD Foundation, "AMD Changing the Game," AMD Corporation, www.amd.com/us/
    aboutamd/corporate-information/corporate-responsibility/community/changing-the-
    game/Pages/information.aspx.
45. Allyson Peerman, interview with author, Austin, TX, July 3, 2012.
46. AMD Corporation, "AMD Foundation Named Computerworld Honors Laureate," press
    release, April 25, 2012, www.amd.com/us/press-releases/Pages/computerworld-honors-
    2012apr25.aspx.
47. Peerman interview.
48. Allyson Peerman, e-mail to author, September 28, 2012.
49. Ibid.
50. Ibid.
51. Jeanne Gustafson, "Microsoft Employee Connects High Tech Professionals with Local
    Students," *Redmond Patch*, October 10, 2011.
52. Rubaiyat Khan, e-mail to author, October 1, 2012.
53. Gustafson, "Microsoft Employee Connects High Tech Professionals."
54. Nick Wingfield, "Fostering Tech Talent in Schools," *New York Times*, September 30, 2012.
55. Todd Bishop, "Geek of the Week: Kevin Wang," *Geekwire*, July 27, 2012, www.geekwire
    .com/?s=Geek+of+the+Week%3A+Kevin+Wang&_opt_paget=GeekWire+-+Dispatches+
    from+the+Digital+Frontier.
56. Colin Miller, e-mail to author, October 1, 2012.
57. Michael Hawker, e-mail to author, October 1, 2012.
58. Cecil Sidwell, e-mail to author, October 2, 2012.
59. Akhtar Badshah, "Technology Education and Literacy in Schools Program Grows by Leaps
    and Bounds," September 27, 2011, http://blogs.technet.com/b/microsoft_on_the_issues/
    archive/2011/09/27/technology-education-and-literacy-in-schools-program-grows-by-
    leaps-and-bounds.aspx.
60. Nick Wingfield, "Fostering Tech Talent in Schools," *New York Times*, September 30, 2012.
61. Isaac Wilson, e-mail to author, October 2, 2012.
62. Digits, "STEM Ambassadors," www.digits.us.com.
63. Digits, "Digits Volunteers Find Creative Ways to Open Student Minds to Math and Science,"
    August 28, 2012, http://digits.us.com/2012/08/digits-volunteers-find-creative-ways-to-
    open-student-minds-to-math-and-science.
64. Ibid.
65. Ibid.
66. Change the Equation, "Design Principles for Effective STEM Philanthropy," July 18, 2011,
    http://changetheequation.org/design-principles-effective-stem-philanthropy.
67. Ibid.

CHAPTER 15

The Pace of Ark Building Quickens

When everything seems to be going against you, remember the airplane takes off against the wind, not with it.

—HENRY FORD

This chapter covers only the first three years of the 2010–2019 decade, but the projected pace of innovative solutions created by businesses and non-profits is already 50 percent greater than in the 2000–2009 decade. The seriousness of the skills gap is becoming apparent to more business and technology executives, and the solutions are becoming ever more creative.

2010: The Broadcom MASTERS

As I was writing this book, I would occasionally head to our cottage on Cape Cod for several days of thinking and writing. It was during one of those trips, when I was gathering information on why middle school students perceive math and science to be hard subjects, that I came across a *Huffington Post* article written by Paula Golden. She is the executive director of the Broadcom Foundation, the philanthropic arm of the Broadcom Corporation, a Fortune 500 semiconductor corporation focused on the wired and wireless markets. And she is a very enthusiastic STEM advocate!

Golden made a strong case in her column that if our education system made math and science applicable to the lives of young students, then math and science wouldn't be perceived to be so hard. Several weeks earlier I had met Ken Venner, the former CIO of Broadcom, so I wrote him a note asking for an e-mail introduction to Paula Golden, and the following week Paula and I talked on the phone.

Moments into our conversation, Paula told me about the Broadcom MASTERS (Math, Applied Science, Technology, and Engineering for Rising Stars). The program, established by Broadcom's senior leadership team and developed by Golden, is produced annually in partnership with the Society for Science & the Public, one of the oldest nonprofit organizations in the United States dedicated to public engagement in science education, and it aims at inspiring middle school students to consider STEM careers.

I really like the Broadcom MASTERS because it targets the critical middle school years of sixth, seventh, and eighth grades. My work with Tech Corps and other nonprofit groups has convinced me, beyond a shadow of a doubt, that those three years are the most important in getting a child, particularly girls, interested in STEM subjects and careers. Golden wrote that "by the time a student reaches middle school, he or she has mastered the fundamentals of basic math and science and communication and is asking, 'What do I want to be when I grow up?'"[1]

Created in 2010 and launched in 2011, the Broadcom MASTERS is a yearlong competition that begins in the spring semester, when sixth, seventh, and eighth graders compete in local science fairs managed by the Society for Science & the Public. From those local events, the top 10 percent of the participants, about 6,000 students nationwide, are eligible to nominate their project for that year's Broadcom MASTERS competition.

From that applicant pool, the Broadcom MASTERS judges (a panel of scientists and engineers) select as semifinalists the 300 student projects that demonstrate mastery of STEM and the ability to innovate. Then from that group, the judges pick the top 30 student projects as finalists, and those teams win an all-expenses-paid trip to Washington, D.C. At the finals, the first-place project wins $25,000, the second-place project earns $10,000, and the third-place project gets $5,000.

I wanted to know what the student participants learned from the Broadcom MASTERS. Paula Golden sent me the following comments.

One student said, "I enjoyed seeing what other students were researching from around the United States. And I was forced to improve my communication skills [all finalists must present their project to the judges in Washington]. I felt honored to be part of the Broadcom MASTERS and am now considering another science project . . . even though it is not required!"[2]

Another student said, "I learned that working in a group is important; if you are not cooperative, you will not be successful. I also learned countless scientific concepts. Finally, I learned there are a lot of other smart kids out there and I have many opportunities available in my future. It was an experience I will remember for a lifetime."[3]

Their experiences in the Broadcom MASTERS affected their ideas about their future careers. One student said, "It continued to fuel my interest in studying science in high school." For another student, it was a life-changing experience: "From participating in the MASTERS, my ideas about going into a scientific field of work were cemented. It sealed the deal for me!"[4]

Do you know a group of talented middle school students interested in showing off their science and technology skills? Log on to www .societyforscience.org and learn how your community can enter a team in the next Broadcom MASTERS competition.

2011: CA Technologies and the Sesame Workshop

I had just completed a meeting with CA Technologies executives at their Manhattan headquarters. Running late for my next meeting on the other side of town, I quickly entered the elevator, where I met a well-dressed man with the shiniest shoes I have ever seen in all my years in business. I introduced myself and complimented the executive on his shiny shoes, and he said, "Oh, I was just at a filming with Grover at the Sesame Workshop."

I have done a lot of video gigs in my career, but never at the Sesame Workshop and never with Grover, so I asked him, "What were you doing there?" His response floored me: "Oh, we were filming a segment to announce our STEM initiative with the Sesame Workshop."

His name was Andrew Wittman, and he is the affable chief marketing officer (CMO) for CA Technologies. I asked him for his business card and followed up several days later with his team about the partnership between CA Technologies and the Sesame Workshop.

CA Technologies is a global IT management solution provider. The Sesame Workshop is most well-known for its feature program, *Sesame Street*, which was launched on National Education Television in November 1969. A primary goal of the program has always been to utilize the medium of television, which easily holds the attention of children under the age of six, for educational rather than merely entertainment purposes.

For me, the amazing aspect of the partnership between CA Technologies and the Sesame Workshop is that it is utilizing the power of television to reach and excite tens of millions of young Americans under the age of six, many of whom live in low-income areas, about the wonders of STEM.

So how did Grover, Cookie Monster, Bert, and Ernie begin a partnership with CA Technologies?

Like my meeting with Andrew Wittman in the Manhattan elevator, it was a chance occurrence. In June 2011, Erica Christensen, the senior director of community affairs for CA Technologies, who also manages the firm's global philanthropic efforts, was attending the inaugural Clinton Global Initiative meeting in America. At a session on STEM topics, Christensen met Lewis Bernstein, the executive vice president of education, research, and outreach for Sesame Workshop.

Sesame Street had developed a STEM curriculum but was looking for a partner that could help it to enhance the content and spread the program. Bernstein and Christensen continued their conversation over lunch and, according to Christensen, "found common ground over our shared desire to spark early interest in STEM education by developing engaging content for children, caregivers, and educators to bring STEM learning into young children's everyday lives."[5] And this starts at age three!

The CA Technologies and Sesame Workshop partnership will include (1) a free online STEM hub with educational resources for children ages three to five, (2) STEM videos, interactive games, and hands-on lesson plans for teachers, parents, and caregivers, and (3) promotional videos that will encourage families to make STEM learning part of their everyday lives.

Wittman will serve as the executive champion of the partnership, and I asked him to tell me what the $1 million Sesame Workshop partnership means to CA Technologies. "We are proud to partner with Sesame Workshop and support the development of creative and interactive programs that will engage children (ages 3–5) in STEM learning. We aim to help young people discover an interest in technology at an early age and consider educational opportunities and careers in technical fields."[6]

Look for Grover and the rest of the *Sesame Street* cast to start talking geek in June 2013, when the CA Technologies and Sesame Workshop initiative launches across America.

2011: IBM's P-TECH

According to a September 2012 Georgetown University Public Policy Institute study, by 2020, 66 percent of *all* jobs in the United States will require postsecondary education and training. Table 15.1 shows the trend.[7]

Table 15.1   Percentage of U.S. Jobs That Require Postsecondary Education

Year	Percentage of U.S. Jobs
1973	28%
1992	56%
2010	59%
2020	66%

If you ask the average American, "What's the goal of a person's education?," most would reply, "College." Yet according to a 2011 U.S. Census Bureau report, only 30 percent of Americans are college educated. It may surprise some, but that's the highest level ever reported by the Census Bureau.[8]

With 66 percent of future jobs in America requiring postsecondary education and only 30 percent of Americans attending college, Stan Litow had an idea. Litow, the president of IBM's International Foundation, the former deputy chancellor of schools in New York City, and the IBM executive who launched Transition to Teaching (see Chapter 14), decided to leverage his experience as a former education administrator. He wanted to create an innovative program aimed at increasing the percentage of American workers who enroll in a postsecondary degree program.

Why is going beyond twelfth grade so important? The IBM Civic Enterprises team studied that issue and released a report that found there are "29 million middle-class jobs in the United States. Jobs that require more education and training than a high school diploma—but less than a bachelor's degree. Jobs that, on average, pay $35,000 or more per year with nearly 10 million jobs paying more than $50,000 annually and 3.6 million jobs paying more than $75,000."[9]

So what was Litow's big idea? In an interview with the *Wall Street Journal*, he said that IBM had "given up checkbook philanthropy" and "was looking for scalable strategies that can be game changers."[10] That's why IBM Corporation went to the administrators of the New York City public school system and proposed an idea to create a radically new kind of school: a unique approach to secondary and postsecondary schooling that began in ninth grade, just like most high schools across America, but completed its work in fourteenth grade.

The city embraced this innovative approach, and in September 2011, Pathways in Technology Early College High School, or P-TECH, was opened as an experimental high school inside the Paul Robeson High School in the Crown Heights section of Brooklyn, New York.

The September 2011 class at P-TECH enrolled 104 ninth graders, and every year through 2016, another class will be added. Every P-TECH student is paired with a mentor from IBM, whose employees donated time to create the P-TECH curriculum.

The New York City College of Technology at the City University of New York is a core partner in P-TECH, and upon graduation from fourteenth grade, the students are awarded an associate degree in applied science, in either computer science technology or electromechanical engineering technology.

Another bonus of graduation from P-Tech was described by New York City Mayor Michael Bloomberg, who said at a ceremony launching P-TECH,

"When they graduate from grade 14 with an Associate's Degree, and a high school degree, they will be 'first in line' for a job with IBM and a ticket to the middle class."[11] P-TECH is working.

Shilpa Menezes, an IBM program manager, said, "When I registered to be an IBM mentor . . . , I honestly didn't know what to expect. . . . I saw mentoring not just as an opportunity for P-TECH's ninth graders, but for me, too. . . . Becoming a mentor seemed like just the right opportunity for me to give back to a younger generation as they moved through life's unexpected, challenging and very exciting changes. I wanted to help young people see these changes as opportunities for growth and success."[12]

What do mentors do? Shilpa and her protégé, a young woman named Cierra, worked on "numerous online and in-person projects together. Cierra is a strong and willful young lady with dreams to make it big in life, and I was initially apprehensive about being able to meet her expectations. . . . I have tried in my own way to make a difference to Cierra, and along the way I've learned from her, too."[13]

P-TECH is also a hit with the kids. A student named Tahmel said this about the P-TECH program:

> I had, or still have, many flaws in my education. None of my former schools focused on the weaknesses of the students like P-TECH does. For example, as I entered P-TECH, I was struggling a lot with math. It was impossible for me to get anything done. I hoped P-TECH would change my horrible habits in math. And that's exactly what it did. I have a long way to go before I can call myself a good mathematician, but I am very impressed with P-TECH, and I really appreciate the math help I am receiving.[14]

Monesia McKnight, a tenth-grade student at P-TECH, is thrilled with the program. Sitting in a P-TECH classroom waiting for her Introduction to Computer Systems class—a class taught by a college professor—to begin, McKnight shared with a *New York Times* reporter, "How great is that?"[15]

P-TECH is starting to spread. Chicago opened a P-TECH school in September 2012. And other cities, most notably Atlanta and Baltimore, plan P-Tech schools in the future. State education administrators in Maine, Missouri, Massachusetts, Tennessee, and North Carolina are also researching a launch.

News of P-TECH has even traveled around the world. Isagani Cruz, writing in the *Philippine Star*, said, "P-TECH has proven in only one year that we can change the system if we have the will."[16]

If you have the will, or the interest, to learn more about this incredibly innovative education model, log on to www.ptechnyc.org.

2012: Udacity

In the summer of 2011, Sebastian Thrun and Peter Norvig, two of the world's best-known artificial intelligence experts and both professors at Stanford University in California, had an idea: they wanted to create an online course on artificial intelligence and offer it for free "to extend technology knowledge and skills beyond this elite campus to the entire world."[17]

Although the students who registered for the Stanford course would not get Stanford grades or credit, "they will be ranked in comparison to the work of other online students and will receive a 'statement of accomplishment.'" Thrun and Norvig's vision was "to change the world by bringing education to places that can't be reached today."[18]

Both professors were strongly influenced by Salman Khan's Khan Academy model, founded in 2006 (see Chapter 14).

Thrun and Norvig were uncertain how many students would register for their free course. Thrun started to build awareness of their fall effort during a summer academic trip through Europe, but only 80 students had signed up. But enrollment skyrocketed to 160,000 individuals after Carol Hamilton, the executive director of the Association for the Advancement of Artificial Intelligence, widely distributed Thrun's e-mail describing the course.[19]

Overwhelmed by the response to the artificial intelligence course, Thrun and Norvig created a new company called Udacity, and on February 20, 2012, they launched two additional courses: Computer Science 101: Building a Search Engine and Computer Science 373: Programming a Robotic Car. By early 2013, Udacity will offer 18 free online courses taught by 19 university professors from around the world.

Udacity's website describes the company's business this way: "Udacity is a totally new kind of learning experience. You begin by solving challenging problems and pursuing [a]udacious projects with world-renowned university instructors, not by watching long, boring lectures. At Udacity, we put you, the student, at the center of the universe."[20]

The course format consists of video lectures, integrated quizzes, and homework designed to promote earning by doing. Upon course completion, the students receive a certificate indicating their level of achievement.

There are four components of Udacity's business model: (1) take any of the 18 courses for free, (2) join the Udacity community of hundreds of thousands of other students and professors, again for no fee, (3) optionally certify skills online or at Udacity testing centers (here there is a fee), and (4) have your resume distributed to one of 20 Udacity partners for no fee. Charles River Ventures and Peter Thrun are the principal financial supporters to date.[21]

In my life and my business, I live by this saying: "Three of anything makes a trend." The Khan Academy (2006), the MIT and Harvard edX physics course (2012), and Udacity (2012) make an online education trend!

When Thrun was building his first online computer science course, critics of the model believed it would be detrimental to the high-fee course model traditionally deployed at Stanford University. Thrun dismissed that criticism by claiming, "I am much more interested in bringing Stanford to the world. I see the developing world having colossal educational needs."[22]

Hal Abelson, an MIT computer scientist who experimented with online courses in 2002 at MIT, said "The Stanford course showed how rapidly the online world is evolving. The idea that you could put up open content at all was risky 10 years ago. Now the question is how do you move into something that is more interactive and collaborative, and we will see lots and lots of models over the next four to five years."[23]

Richard A. DeMillo, the director of the Center for 21st-Century Universities at Georgia Tech, in commenting on the online education market, which is often referred to as MOOCs (massive, open, online courses), said, "This is the tsunami. . . . It's all so new that everyone's feeling their way around, but the potential upside for this experiment is so big that it's hard for me to imagine any large research university that wouldn't want to be involved."[24]

Five years ago, people used to go to work in a physical building. That was until the smartphone phenomenon occurred. Now work is wherever you are. I am convinced that the same shift will happen to school because of online offerings like Udacity, the Khan Academy, and ed/X, creatively destroying the old model of education with a new one in which learning doesn't take place only from 8 a.m. to 2:30 p.m. each day.

School will be wherever a child has an Internet connection. Online learning models will offer the best teachers and instructors to students around the world.

2012: CA Technologies: Tech Girls Rock

I was sitting in the lobby of the Intercontinental Hotel in Prague reviewing my notes for a midmorning speech I was slated to give at a CA Technologies European IT Summit meeting. As I finished making a last-minute change to a slide, I looked up and saw George Fischer, the executive vice president of worldwide sales for CA Technologies, standing alone, waiting for a breakfast guest. I have known George for several years, so I set aside my presentation material and walked over to say hello.

We talked for several minutes, and during the conversation I mentioned to George that I was working on a book about science and math. He told me about Tech Girls Rock, a newly launched STEM initiative he led at

CA Technologies in partnership with the Boys & Girls Clubs of America, a group that CA Technologies had worked with since 2005.

Later that day I logged on to the site and discovered that Tech Girls Rock is a program "to empower girls to explore a future in technology and is aimed at helping girls to discover and cultivate an interest in technology, and tech-related educational and career opportunities."[25]

The CA Technologies commitment is twofold: (1) a $1 million contribution to Boys & Girls Clubs chapters all across the country, and (2) volunteer efforts by CA Technologies employees as moderators, panelists, and group leaders at technology Boys & Girls Clubs workshops held in cities like Boston, New York, Plano (Texas), Chicago, and San Francisco since the launch announcement in late 2011.

According to the research, one big reason that girls shy away from pursuing STEM careers is that they have no mentor, no role model, of a successful woman in the tech field.

That's a key objective of the CA Technologies and Boys & Girls Clubs Workshops. Successful female tech professionals from CA Technologies and other companies participate on panels that address IT trends and lead hands-on projects that give girls and young women a better understanding of what it means to be a female tech professional.

Here's what a reporter from a San Francisco newspaper found. A Boys & Girls Clubs member who attended the event claimed that she "was a bit hesitant about the event at first. It [technology] is something that a lot of girls are shy about. They never have had the tools to fix a computer, so they don't get the chance to do that. Then I realized, I'm not alone . . . other girls haven't done this stuff either, but we get to work as a team at Tech Girls Rock."[26]

At each of the 2012 Tech Girls Rock events, *My Club, My Life*, the online reporting arm of the Boys & Girls Clubs of America, assigned a young reporter to cover the event. The reporter who participated in the East Harlem, New York, event wrote that "the girls who attended Tech Girls Rock shared their dreams of becoming dancers, doctors, and veterinarians. When the CA Technologies explained how technology can be incorporated into all of these fields, their eyes widened. They had never heard of being able to use computers to keep digital records or machinery that is able to use physics to structure a perfect dance leap!"[27]

And at the South San Francisco event, a participant said this:

> When we went to the break-out groups, we got a chance to talk about what we wanted to be when we grow up. Our CA Technologies group leader said that we all needed to go to school and graduate with a degree to reach our goals. One of my goals is to go to college and be an author like J. K. Rowling. Attending Tech Girls Rock has inspired me to learn more about technology—such as Photoshop and how technology is needed for architecture jobs. I hope to attend the event next year.[28]

Jim Clark, the president of Boys & Girls Clubs of America, said the following:

> The Tech Girls Rock initiative made possible by CA Technologies equips Boys & Girls Clubs across the country with the capacity to focus tech skills development specifically for young women, ensuring a future workforce that is job ready and destined for a great future in the technology field. Tech Girls Rock is a key part of the Boys & Girls Clubs of America's strategy to ensure all members graduate from high school on time and are equipped for a postsecondary education needed for a 21st-century career.[29]

George Fischer is pleased with the results of the first year: "It is important for CA Technologies to support and foster an interest in IT early on for young women. So many studies underscore the need for more diversity and women leaders in the information technology industry. By leveraging the power of technology, Tech Girls Rock aims to continue to engage and excite these girls about the many opportunities and benefits a career in IT can provide."[30]

CA Technologies employees are also satisfied with their participation in Tech Girls Rock. Debra Danielson, the distinguished engineer and senior vice president of merger and acquisition strategy at CA Technologies, shared with me, "What a treat it was to meet with young women from Boys & Girls Clubs of America in our Tech Girls Rock workshops. It was fun to talk with them and answer their questions about careers in technology. I hope we sparked some interest and encouraged these bright and motivated youngsters to consider careers in the IT sector."[31]

Judy Kruntorad, vice president of quality assurance at CA Technologies, added, "I have been working for CA Technologies for over 20 years, and in the software industry for almost 40, and truly enjoyed helping to encourage the next generation of women technology leaders during our workshops."[32]

Tech Girls Rock does exactly that. The more than 1,000 young women from around the country who have participated in the program can vouch for that. More events are planned for Bellevue (Washington), Scottsdale (Arizona), and Tampa (Florida) in 2013.

2012: Microsoft's Teach.org

The teacher recruitment pipeline in America is too one-sided. That is, our country recruits more than 85 percent of new teachers from new college and university graduates. Many of them possess education degrees but have no degree in science, technology, engineering, or math. That has to change.

As millions of currently employed teachers retire or plan to retire in the coming decade, America needs to replenish the supply of math and science teachers from the pool of IT workers who do have STEM undergraduate, or more advanced, degrees.

The Department of Education held an auction in late 2011 for the future maintenance and operation of its Teach.gov web site, whose content is aimed at inspiring visitors to the site to "make a difference" and choose a career path to teaching. The Microsoft Corporation was the winner, and Teach.gov is now Teach.org, a rich repository of information, personal stories, and data designed to answer the seminal question "Why teach?"

Teach.org lists three primary reasons that IT workers, seeking perhaps a second career as a science or math teacher, should consider the profession of teaching:

1. **Make an impact**. Teachers influence lives, now and into the future. As a teacher, you have the unique ability to promote opportunity for all children. Change the future. Teach.
2. **Be a leader**. In many professions, it takes months, even years, to get a leadership position. When you become a teacher, you become a leader the very first day you stand in front of the class. Better yet, you help build leadership skills within your students.
3. **Build a career**. Teaching is not just a job. It's a profession that offers the ability to grow and advance. You won't be limited to the same classroom year after year.[33]

America's future depends on replacing retiring baby boomer teachers with a new generation of teachers who stand before middle school and high school math and science classes with the added minimum credibility of having an undergraduate degree in science, technology, engineering, or math.

Kudos to the Microsoft Corporation for hosting Teach.org. Thinking about a second career as a teacher? Visit www.teach.org.

2012: The Dell Education Challenge

In September 2012, Dell Inc. launched the first Dell Education Challenge as part of the larger Dell Social Innovation Challenge program. The Dell Education Challenge (www.dellchallenge.org/k12) will focus on unearthing ideas that provide solutions to tough education problems.

Dell proposes solutions that "can address in-school or out-of-school learning, the student, educator, or the learning environment, or systemic issues with infrastructure, policy, or administration." It will "act as an educational innovation community where students, academia, K–12 educators, and fellow students from around the globe can network, share best practices, and inspire each other to create educational change."[34]

The challenge also comes with perks, such as $30,000 in awards, and all finalists are flown to Austin, Texas, to attend Dell World, where teams will compete live in front of a panel of judges to tout the educational innovation merits of their project. With technology dramatically changing the educational landscape, competitions like the Dell Education Challenge have the distinct possibility to unearth the next big thing in education.

The model for the Dell Education Challenge follows closely the model used in the Dell Social Innovation Challenge competition, which has a much broader scope than just education. It focuses on identifying and supporting young social innovators who submit project applications aimed at solving some of the most pressing problems in the world.

The 2012 winners in the Dell Social Innovation Challenge were a team from Tulane University that built the Humanure Power Project, which dramatically improved sanitation and electricity infrastructure in scores of villages in India; a collaborative team of students from Duke University, MIT, and Berkeley that formed a company called Nanoly, which increases the accessibility and affordability of vaccines for developing countries; students at Arizona State University who created a project called 33 Buckets, which provides a girls' school in Bangladesh with arsenic-free water; and Dhaka University students in Bangladesh who developed E-Education in Five Continents, which provides video lecture classes to poor rural students in Bangladesh, the West Bank and Gaza, and Rwanda.

This is incredible stuff, and it's a vivid example to Americans that young people in all parts of the world are highly creative, highly innovative, and well versed in science and math.

2012: The Girl Scouts of America's *Generation STEM: What Girls Say about Science, Technology, Engineering, and Math*

Earlier you read about the American Association of University Women report that claimed that positive, early reinforcement from parents and teachers has a powerful effect on a young girl's perception of her science and math skills and on whether she will aspire to a career demanding strong math and science skills.

The leaders of the Girl Scouts of America, concerned about the low representation of women in all levels of STEM careers, directed the organization's research division to conduct a study to discover what girls like about

STEM. The study, *Generation STEM: What Girls Say about Science, Technology, Engineering, and Math*, reported five broad findings.[35]

1. Most Girls Like STEM

A strong gender stereotype, particularly in the middle school grades of six through eight, is that math and science, as subjects, appeal more to boys than girls. But the Girl Scout Research Institute Study refutes this stereotype by reporting that 74 percent of teenage girls (ages 13–17) are interested in STEM.

2. The Creative and Problem-Solving Aspects of STEM Interest Girls

Girls interested in STEM like to understand how things work (87 percent), solve puzzles and problems (85 percent), do hands-on science projects (83 percent), ask questions about how things work (80 percent), and understand how the natural world works (79 percent). That's an amazing list! And I guarantee you that if you read that list to a colleague and asked how many girls like to do those things, the percentages in the responses would be much lower.

3. Certain Influences Set STEM Girls Apart

The Girl Scouts categorized the respondents by girls who said they were "somewhat or very" interested in STEM—STEM girls—and those who were not.

The Girl Scout report concluded that the girls who were interested in STEM fields were higher achievers, had stronger support systems, and had a higher assessment of their own intelligence (71 percent said they were smarter than other girls their age). Interesting to note is that STEM girls were less likely to say they would be famous one day.

STEM girls see themselves on the same playing field as boys, with 97 percent agreeing that "whatever boys can do, girls can do." They attribute their success in math and science to hard work and determination.

Parents (particularly fathers who are very interested in STEM), career role models, and museum visits set STEM girls apart from those not interested in science and math. About 70 percent claimed their "parents try hard to make sure I am exposed to many career options," 66 percent said that visiting science or technology museums was influential to them, and 66 percent also claimed that knowing someone in a STEM career shaped their interest level in STEM.

4. A Gap Exists between STEM Interest and Career Choice

The report stated that even though 81 percent of girls interested in STEM expressed an interest in pursuing a career in a STEM field—specifically in engineering (32 percent), the physical sciences (57 percent), math (31 percent), computer science or IT (27 percent), or software development (25 percent)—only 13 percent said that a career in those fields was their first choice.

When the study asked why, 57 percent said it was basically too early in their lives to consider a career in STEM, 47 percent said they were uncomfortable being the only girl in a group or a class (presumably in middle school or high school), and 57 percent said they'd have to work harder than a man in a STEM field just to be taken seriously.

5. The Story Differs for African American and Hispanic Girls

African American (76 percent) and Hispanic (74 percent) girls expressed a slightly higher interest in STEM careers than non-Hispanic Caucasian (73 percent) girls. But the study pointed to key challenges for African American and Hispanic girls: "They have less exposure to STEM, less adult support for pursuing STEM fields, lower academic achievement, and a greater awareness of gender barriers in STEM professions. However, their confidence and ability to overcome obstacles are high, pointing to the strong role of individual characteristics in STEM interest and perceived ability in these subjects."

Other key obstacles for African American and Hispanic girls are the following: (1) only 48 percent of African American girls and 52 percent of Hispanic girls know someone in a STEM career, compared to 61 percent of non-Hispanic Caucasian girls, and (2) only 54 percent of African American and Hispanic girls said they could rely on their parents for information on career choices, compared to 70 percent of non-Hispanic Caucasian girls.

As I was reviewing the Girl Scouts of America STEM report, I recalled a column I had written in *CIO* magazine in 2008 about the efforts of Sophie Vandebroek, the chief technology officer for Xerox and a recent inductee into the Women in Technology International's Hall of Fame, to build a more gender-balanced engineering workforce at Xerox.

According to the U.S. Department of Labor, only 17 percent of high-tech jobs are held by women.[36] Xerox has focused on creating a gender-balanced workforce for decades and has achieved this in its sales and marketing divisions. However, in the engineering and research divisions, women are still very much underrepresented. A decade ago, when Vandebroek began focusing on achieving a gender-balanced workforce in engineering at Xerox, only 15 percent of the company's new engineering employees were women. Now, more than 40 percent of new engineering employees at Xerox are women.[37]

For my CIO column, I asked Vandebroek to explain the company's commitment to building a more gender-balanced workforce. She told me,

"An inclusive environment allows Xerox to be more innovative. Our global customers are from diverse cultures worldwide, and many are women. It's essential to have the female and global perspective represented in our products and services."[38]

Several years ago, Vandebroek elaborated on the topic of gender balance in a speech at a computing conference: "Having a diverse engineering team enables exceptional performance; in addition to outstanding technical skills, women engineers naturally bring skills such as multitasking and relationship building. In order to get a project done you need to work with people across disciplines—technology, marketing, finance, manufacturing, sales, etc.—and creating what I call a strong human fabric is critical to getting things done."[39]

In the same speech, Vandebroek shared this reality check with the conference attendees:

Bottom line: the number of jobs in the U.S. economy that require engineering and science degrees is growing, the number of people prepared to fill these jobs is shrinking, the availability of qualified people immigrating from other countries is flat and expected to decline as their home countries become more competitive and the U.S. becomes more restrictive.

The shrinking pool of young people with the skills to innovate won't impact our standard of living overnight, but it's like an invisible cancer that is slowly growing larger. We must start to address the issue now. It's a shared responsibility of U.S. business, government, and education . . . but one thing is sure, we must find ways to encourage more girls to become scientists and engineers. This provides a huge opportunity. We must start getting girls interested in technology at a very young age before they enter middle school. We must discredit the scientist and engineer stereotypes. We must shelter girls from a stereotypical world of Barbie dolls, pink and glitter.[40]

Vandebroek is absolutely right—not only about gender balancing in the tech workforce but also about the broader concern about improving STEM education as a "shared responsibility of U.S. business, government, and education."

News Alert: More Arks Needed!

I hope you have been informed and inspired by the many efforts of businesses and nonprofit organizations outlined in this section of the book.

The warnings in Part Two and the "arks" in Part Three have been presented in chronological order to determine whether there was any correlation between the warnings and the solutions. Table 15.2 provides a summary.

Table 15.2   The Apparent Need for More Solutions

Decade	Warnings	"Arks"
1960s	1	2
1970s	3	0
1980s	9	1
1990s	13	5
2000–2009	39	15
2010–2019*	53	25

* Warnings and "arks" for the full decade of 2010–2019 are prorated for a full decade based on the actual number discovered from 2010 to 2012.

I have several observations about these numbers. First, for the 30 years from 1960 through 1989, there was relatively little interest in STEM education in America; there were neither warnings that something was off track nor solutions proposed by groups. Warnings did outnumber "arks," in my highly unscientific but thorough investigation, by roughly a four-to-one margin during this period.

Second, in 2000–2009 there was a significant increase in the number of both warnings and arks, particularly starting in the middle part of the decade. There could be several reasons for this. The relatively mild recession of 2001–2002 put work skills in the spotlight. So did the high-profile No Child Left Behind Act, which sought to remedy the eroding scores of American students in math and science tests. The business community also became a more active vocal critic of the U.S. public education system at this time.

Perhaps the most profound observation I can offer, however, is this: based on just the first three years of the decade 2010–2019, both warnings and arks are increasing at a record pace, suggesting to me that the topic of STEM education in our country will only increase in the future. I predict that STEM warnings in the coming seven years will outnumber arks by nearly two to one.

So what does all this mean? Well, recalling Louis V. Gerstner Jr.'s comment (see Chapter 11), it is still raining. Theodor Geisel, the American writer better known as Dr. Seuss, wrote in the children's classic *The Butter Battle Book*, "Sometimes the questions are complicated and the answers are simple." For the United States to close the science and math technology skills gap, the questions that must be asked remain complicated. But the answer is simple: Our country needs more arks—lots of them. America needs the entire community of 5 million IT executives in the United States to start building more of them, right now.

Notes

1. Paula Golden, "Keep Math and Science Alive in Middle School," *Huffington Post*, October 4, 2011, www.huffingtonpost.com/paula-golden/keep-science-and-math-ali_b_993380.html.
2. Paula Golden, e-mail to author, August 19, 2012.
3. Ibid.
4. Ibid.
5. Erica Christensen, e-mail to author, August 28, 2012.
6. Andrew Wittman, e-mail to author, October 8, 2012.
7. Georgetown University Public Policy Institute, *Career and Technical Education: Five Ways That Pay* (Washington, DC: Georgetown University Public Policy Institute, 2012).
8. U. S. Census Bureau, March 2011.
9. IBM Corporation, *Enterprising Pathways: Toward a National Plan of Action for Career and Technical Education*, September 2012.
10. Stephanie Banchero, "New York Teams Up with IBM to Reboot a High School," *Wall Street Journal*, August 1, 2011.
11. P-TECH High School, www.ptechnyc.org.
12. Shilpa Menezes, "The Joy of Mentoring," *Citizens IBM Blog*, June 5, 2012, http://citizenibm.com/2012/06/the-joy-of-mentoring.html.
13. Ibid.
14. Ibid.
15. Al Baker, "At Technology School, Goal Isn't to Finish in Four Years," *New York Times*, October 22, 2012.
16. Isagani Cruz, "Mini Critique," *Philippine Star*, September 27, 2012.
17. John Markoff, "Virtual and Artificial, but 58,000 Want Course," *New York Times*, August 15, 2011.
18. Ibid.
19. Ibid.
20. "Udacity: 21st Century University," www.udacity.com.
21. "Udacity: The Four Elements of Udacity," www.udacity.com.
22. Markoff, "Virtual and Artificial."
23. Ibid.
24. Tamar Lewin, "Universities Reshaping Education on the Web," *New York Times*, July 17, 2012.
25. Boys & Girls Clubs of America and CA Technologies, "Boys and Girls Clubs of America and CA Technologies Launch Tech Girls Rock," press release, December 6, 2011.
26. Jill Pantozzi, "Tech Girls Rock Initiative Launched," *San Francisco Patch*, April 5, 2012.
27. Corinne H., "Tech Girls Rock: NYC & Boston," My Club, My Life, February 21, 2012, www.myclubmylife.com/Arts_Tech/Pages/tech-girls-rock-nyc-boston.aspx.
28. Rebecca A., and Amanda B., "Tech Girls Rock: South San Fran," My Club, My Life, www.myclubmylife.com/Arts_Tech/Pages/tech-girls-rock-san-fran.aspx.
29. Boys & Girls Clubs and CA Technologies, "Boys and Girls Clubs of America and CA Technologies Launch Tech Girls Rock."
30. George Fischer, e-mail to author, September 26, 2012.
31. Debra Danielson, e-mail to author, October 9, 2012.
32. Judy Kruntorad, e-mail to author, October 9, 2012.
33. Microsoft Corporation, "Why Teach?," n.d., www.teach.org.
34. Dell Inc., "Dell Education Challenge," n.d., www.dellchallenge.org/about/education.
35. Girl Scout Research Institute, *Generation STEM: What Girls Are Saying about Science, Technology, Engineering, and Math*, Girl Scouts of America, 2012, www.girlscouts.org/research/publications/stem/generation_stem_what_girls_say.asp.
36. U.S. Department of Labor, "Facts on Working Women," n.d., www.dol.gov/wb/factsheets/hitech02.html.

37. Sophie Vandebroek, panel, "Elevating the Role of Women: Insight from Women Who Hold Board Positions in the Technology Industry," Grace Hopper Celebration of Women in Computing Conference, Atlanta, September 30, 2010.
38. Gary Beach, "For a Gender-Balanced Workforce, Businesses Should Copy Xerox's Approach," *CIO* magazine, January 30, 2008.
39. Vandebroek, "Innovation and Leadership."
40. Ibid.

EPILOGUE

For What It's Worth

Buffalo Springfield wrote and released "For What It's Worth," one of rock music's most memorable songs, in 1966. The song notably warned about the uncertainty of a sharply divided American society to reach consensus on important issues. The song's title, and message of warning, serves well on several fronts to frame my recommendations on how to address, and solve, the technology skills gap currently confronting the United States. First, for what it's worth, as I mentioned in the book's preface, I am no expert in education. But as a parent whose two children went through the public school system, as a media professional in the information technology business for 30 years where I have listened to CIOs lament that America's schools are not preparing students for the twenty-first-century workforce, and as a founder of one of the country's oldest technology nonprofits serving schools, I have formed opinions—that some readers will embrace and others will criticize— on what needs to be done. Perhaps my best credential for sharing with you my recommendations is this: the thousands of hours of research I have invested in trying to piece together a rational story about why American schoolchildren have fallen so far behind the rest of the world in math and science, and why it is that the further a U.S. student goes in our public school system the further behind our nation gets.

American students didn't just wake up in 2013 and find themselves in this fix. Since the mid-1960s, there has been something happening to the quality of math and science education in our country. And most of it isn't good. In 1963, SAT math scores for high school students started a decline for the next 13 years. One year later, in 1964, the First International Math Study ushered in five decades of international math and science tests that documented weak performance by American students compared to their peers in both developed and emerging countries around the world. And despite the U.S. Department of Education's 25-year focus on accountability and compulsory testing, two out of three students in the United States in 2013 test less than proficient in either math or science.

Over the past half century, something did become clear: pervasive uncertainty about what to do and the inability of a country to have critical stakeholders meet regularly and share ideas on how to systemically improve public education in the United States.

There remains consensus on how to reverse the prolonged slide in math and science education in the United States. For nearly five decades, lots of smart people and lots of organizations offered lots of opinions. These opinions differed, had short shelf-lives, were quickly forgotten, and were rarely acted upon.

Some people nevertheless believe that the American education system, in general, and how our nation teaches math and science to 49,266,000 students, in particular, is doing fine. In my talks to IT executives, I respond to such logic with this comment: "I am glad you are satisfied with the status quo. But I have a challenge for you. I want you to imagine an empty whiteboard. If you had the opportunity to design the ideal American education system for the twenty-first century, would your plan look like our current system or not?" Most respond that their plan would be dramatically different from our current K–12 system, which was designed for a nineteenth-century manufacturing and agricultural society. And that's my point: even the supporters of the current system admit it needs reform. But no one seems to have the big idea that will show the way forward.

Meanwhile, in that environment of confusion, American school children continue their mediocre performance in math and science.

I am amused with the general public's "the sky is falling" reaction to newly released math and science test scores that plot further mediocre performance by American students in national and international math and science tests. Here's why. It's as though American students once performed well in math and science tests and, for some reason, have started to slip in the last 10 years. Far from it. America has never been on top. Never.

America hasn't even been a close second, which is a national embarrassment for a nation that has increased annual expenditures per pupil 130 percent since 1970, resulting in nearly $600 billion being spent on education in America. (I was careful not to use verb *invested* in the previous sentence.)

But all is not lost. Although American students may have not mastered math or science for more than a century, the United States has taken several significant legislative strides in broadening education opportunities since the end of World War II. The GI Bill helped millions of World War II veterans enter and pay for college. Today, 30 percent of Americans have college degrees, the highest percentage on record. The historic 1965 Elementary and Secondary Education Act, ushered in during President Johnson's Great

Society era, broadened the educational opportunities for millions of under-privileged Americans. And 93 percent of American high school seniors now graduate with a high school diploma; this is also an achievement, up from the abysmal rate of about 40 percent at the conclusion of World War II.

But other than those mostly legislative milestones, America has not focused on *how* to educate, despite a cacophony of calls to restructure the country's nineteenth-century factory-style K–12 education system, which is more designed to promote mediocrity than a meritocracy.

Critics claim that the K–12 public education system is the last bastion of American society that hasn't been reformed since the Civil War. I agree, particularly when you look at other fields. Since the late nineteenth century, computers have evolved from the rudimentary programmable computer built by Charles Babbage to incredibly powerful supercomputers and equally pow-erful tablets. Talking on the phone has evolved from Alexander Graham Bell talking with "Mr. Watson" to billions of people speaking on smartphones. Advances in aviation technology have taken humanity from short flights over the sandy hills of coastal North Carolina to the barren landscape of the moon 240,000 miles away.

But if a child from 1890 was magically transported to a classroom in 2013, that child, unfortunately, would be remarkably comfortable. There would still be one teacher in the room. Students sit in neat rows of seats. And reading, writing, and arithmetic are the primary subjects taught in the classroom.

America needs to do better. America must do better.

The United States responds well when an apparent, identifiable force threatens its survival. But obtuse challenges, like systemically restructuring science and math education, are difficult for the United States to confront, fully understand, and respond to.

Want proof? Thirty years after *A Nation at Risk* warned us that our me-diocre educational performance would be viewed as an act of war if a foreign power had imposed it on us, our nation continues to debate the *same* educa-tion issues. There is just one significant exception: China, rather than Japan, is now the foreign power our country is most concerned about in the twenty-first century.

Improving math and science education in America is not about build-ing a nation of math and science eggheads who outscore their peers in other countries around the world. Just as literacy was the key issue for most of the twentieth century in America, proficiency in math and science is the key issue for the twenty-first century. Building a nation with a population with strong math and science skills is the direct path to three strategic conditions for our country: a more robust economy, a more employable workforce, and a more secure environment for every American.

A nation of citizens with strong math and science skills is the foundation of our country's future. And on that foundation our schools must now teach the "new" skills, the five Cs—critical thinking, collaboration, communication, creativity, and confidence—because the combination of strong math and science skills with the five Cs is our nation's most logical path to innovation and invention in the future. A nation whose citizens excel in the five Cs but not in math and science will be a nation of liberal arts majors. A nation of math and science wizards who can't think critically, communicate, collaborate, create, or be confident will be a nation of machines. The successful countries of the twenty-first century must do *both* well.

My biggest fear for America is this: if the People's Republic of China integrates a five Cs curriculum into its schools before America does, it is game, set, and match for our country!

Sun Tzu, a Chinese general who wrote the military strategy book *The Art of War*, claimed that "strategy without tactics is the slowest route to victory, while tactics without strategy is noise before defeat."

The Top 10 Recommendations for Action

I like Sun Tzu's advice. So here are my top 10 recommendations for what America, and America's IT executives, need to do to ensure the future competitiveness and survival of our country. My first three recommendations frame my strategic ideas, which form the foundation of my recommendations. The remaining seven recommendations are my tactical proposals on how to implement that strategy.

America must do both! I honestly don't know whether they will work. But I do know that even though my list of recommendations is relatively short, it is the result of six years of research on this topic.

I also know that whatever course America decides upon, if in fact our country can make a decision, it will take a very long time for our country to reboot our national education initiatives. It could take at least 30 years, but probably longer. Remember how long it took the United States to embrace W. Edwards Deming's total quality management approach to manufacturing? What I worry about most is whether American society has the collective patience to see this job through to completion.

1. Create a Long-Term National Education Strategy

Most Americans are aware of our nation's fiscal deficit, in which the U.S. government spends more than it takes in every year.

I believe that our country also faces a national education deficit. America spends more than a half trillion dollars each year on public education and gets back little in return. This deficit is less tangible than a fiscal deficit, and it cannot be measured by counting dollars. But it is a deficit nonetheless, and it poses a serious threat to the future economic, employment, and national security future of America. Shirley Jackson, the president of Rensselaer Polytechnic Institute in Troy, New York, might call it the "quiet" deficit.

My first recommendation is therefore the creation of a long-term national education strategy that will deliver a singular result: that the United States will have the most innovative and most inventive workforce in the world by 2030. This workforce must be the best in the world at thinking critically, collaborating with workers (whether they are across the hall or on the other side of the world), and being highly analytic. The workforce must be able to clearly communicate a course of action, be highly creative, and excel at invention and innovation.

At present, no long-term national education strategy document exists. We have laws like the America COMPETES Act, which offers scores of great ideas our nation could do. But the problem with such a legislative provision is that it must be funded, and that depends on the reauthorization whims of Congress. As such, it has no long-term viability.

We also have a national educational technology strategy that calls for applying the advanced technologies in our daily lives to our entire education system. But our country has been doing this since the time the Apple IIE showed up in classrooms across the country in the early 1980s, creating opportunity as well as confusion on how to use the latest advanced technology in our schools. I have often found that plans like this are written by technology companies that have a vested self-interest in selling more stuff to schools.

Here's how I would frame a long-term national education strategy. I would have the president form an independent commission of business leaders, educators, high school and college students, and academic leaders to construct the strategy and report biannually to America on the state of education in the United States. I see the commission's purpose akin to the Erskine-Bowles Commission that made bold recommendations to the nation on how to reduce our national fiscal deficit.

I would fund the commission with large donations from key foundations focused on improving math and science education, specifically, and overall education in general. I also see a huge opportunity to raise tens of millions of dollars from microdonations of one dollar. Education is important to most Americans, most Americans believe our current system needs to be fixed, and most Americans could contribute a buck to the cause.

The commission would be nonpartisan, nonpolitical, and independent. As a courtesy, it would report its findings to the president of the United States, but the commission would not be beholden to the president or any federal, state, or local government agency.

The commission would, however, have strong powers of assessment. Most Americans understand the letter grade system. I would empower the commission to grade—from A to F—the nation's education system in areas like (1) infrastructure, (2) quality of teachers, and (3) quality and relevance of course curricula (with a particular focus on embracing online course curricula).

The commission would issue a biannual report card. It would be easily understood by students, parents, teachers, administrators, politicians, and the media and would keep the vitality of our nation's education system a front-burner issue for all Americans. The biannual report card is critical, because the United States no longer has the time to read another scathing report like *A Nation at Risk* and 30 years later be in the same situation. The commission must keep the health of our nation's public school system front and center, and this report card approach, though elementary, would do just that. The commission might also consider posting the grade letters of every school in America in the school's lobby for all to see, not unlike how boards of health post letter grades in the windows of restaurants.

Because education is a means to an end, the vitality of American society in the future, I would encourage the commission to review broader measures than education. I would task the commission to measure the success of a national education strategy by issuing a quarterly index composed of yearly changes in the following key economic, workforce, innovation, and national security issues: (1) GDP growth, (2) unemployment trends, (3) the high school dropout rate, (4) student proficiency performance in all subjects in the NAEP tests, (5) the number of cyberattacks, (6) America's ranking in the World Economic Forum's *Global Competitiveness Report*, (7) the amount of U.S. government funding in basic research, and (8) the number of triadic patents filed by Americans (a report easily found on the Organization for Economic Cooperation and Development's web site). This quarterly report would be similar in nature to the Conference Board's Leading Economic Indicators Index.

The initial index, which I would call the nation's Leading Education Index, would be tagged to a quantitative value of 100. Subsequent quarterly indexes would be mapped to the 100 value. An index below 100 would suggest our national education strategy weakened in the quarter. A number above 100 would indicate progress.

The quarterly Leading Education Index and the biannual letter graded report card on the country's overall education system would be an easy-to-understand measurement informing Americans whether our nation is making progress toward the goal of having the most skilled workforce and innovative economy in the world by 2030.

2. Determine Accountability for the National Education Strategy

Once a national education strategy is set and the benchmarks are agreed upon, there remains a thorny issue: Who is accountable to see it through? The commission? The president? The secretary of education? Business and academic leaders? The 535 members of Congress? The 50 state governors? The 14,000 school superintendents? The 99,000 public school principals? The 3.5 million teachers? The teacher unions? The tens of millions of parents of school-age children? Or the 49,266,000 million kids themselves?

This is quite a long list of constituencies, each of which currently has an important voice in the direction of public education in our country. These voices, however, often form a national education tower of Babel. So let's winnow through the list to arrive at my recommendation for accountability.

Since the words *education* and *educate* are nowhere to be found in the 4,512 words of the Constitution, and because the Department of Education was founded in 1979 solely as political payback from President Carter to the National Education Association, I recommend downgrading the Department of Education from cabinet status to office status within the Department of Commerce. The new U.S. Office of Education would serve as the nation's primary education statistical gathering unit, a role the unit served in its pre-cabinet incarnation as an office in the Department of Health, Education, and Welfare.

The creation of subcabinet status for the Department of Education serves another important purpose: it removes the president from direct responsibility for educational matters. The problem I have with the current structure is that with the president leading the nation's education efforts, the longest-term national education strategy is only eight years. That's not long enough, and it's even more of a concern when you recall that 64 percent of American presidents have served one term or less.

Some IT executives with whom I have shared this recommendation counter with "Wouldn't that dramatically impact the federal funding for public education in America?"

The answer is not really, because 95 percent of the $583 billion annually spent on public education in the United States is funded from state and local taxpayers, which offers a clue on one group that must be highly accountable for a national education strategy: parents.

What about Congress? What should be the role of the nation's lawmaking body pertaining to a national education strategy? I refer again to the Constitution: nowhere does it stipulate that power be invested in Congress for education. If Congress wants to continue to reauthorize funding bills like the Elementary and Secondary Education Act or the America COMPETES Act, so be it. But I disagree with it, because members of Congress are more interested in the parochial concerns of their respective districts or states.

Look what Congress's last major legislative act, the No Child Left Behind Act, has done for American education. Not much. Congress deserves to have no accountability for a national education strategy.

But the 50 state governors do. They will play key roles in framing and being accountable for a national education strategy. In fact, I recommend that the National Governors Association actually host the first national education strategy meeting where the platforms of the strategy are framed and voted on.

Whether or not that happens, the nation's governors must take education more seriously. Governors currently meet twice a year to discuss a wide range of issues, including education. The governors association also has initiative groups, which meet less frequently. For instance, according to the National Governors Association web site, the education initiative group last met in July 2010. That is not acceptable.

Business leaders must absolutely be held responsible for building and executing a national education strategy. They are definitely interested. Business leaders have a keen sense of the twenty-first-century skills their workforces need, and their voice at the accountability table should be very loud.

As mentioned earlier, parents must be accountable for a national education policy. After all, they pay 95 percent of the nation's education bill. They must do a much better job of identifying children gifted in science and math and, possibly more important, give their children, especially their daughters, the confidence that they can succeed in math and science.

If the parents themselves feel overwhelmed by math and science, they can and should utilize the power of online initiatives like the Khan Academy in their homes to help reinforce math and science skills in their children. The success of a long-term national education strategy, particularly as it pertains to STEM, hinges on parents becoming more proactive in public education. If parents remain on the sidelines, there is no chance that a national education strategy will succeed.

A constituency that must be kept as far away as possible from framing and being accountable for the successful implementation of a long-term national education strategy is teacher unions. Teacher unions, in my opinion, are concerned about two issues: the long-term financial stability of their members, and defining, through collective bargaining contracts, the work rules of our nation's public school teachers. I suggest that the National Education Association, the nation's largest teacher union, adopt the name it originally had at its founding in 1857: the National Teachers Association. The National "Education" Association is a misnomer.

Who really knows who the best math and science teachers are in your school? It's not the department head who observes your child's teacher for less than an hour once per year. Nor is it determined by the results of the infamous No Child Left Behind test. The accountability for determining the best teachers should fall to the parents.

Here's how I would do it. I recommend that all 50 states conduct, in each school district, a Web-based teacher satisfaction survey. The opt-in survey would be conducted among the parents of the children attending a specific school. It would ask a battery of questions aimed to determine who are the best teachers at the school. Another prerequisite for the test is to make the results public. The $583 billion that American taxpayers invest in public education every year earns them the right to know who the best teachers are.

There's one more group, perhaps the most important, that also needs to be accountable for framing and executing a national education strategy: the tens of millions of middle school and high school students.

In Chapter 4 you read about the *Life* magazine series published in March 1958 that depicted Soviet teenagers working much harder at studying math and science, and other subjects, than American teenagers.

Fast-forward to 2013. Students in Japan go to school for 240 days a year. Students in South Korea have a 10-hour school day that begins at 7:30 a.m. and ends at 5:30 p.m. American kids, often more distracted by video games, texting their friends, sports, or talking on phones paid for by their parents, go to school for six hours a day, 180 days a year.

Nothing will work in a national education strategy unless America's middle school and high school students are held directly accountable for their educational outcomes. And here's my idea to do just that.

Peer pressure is a powerful force. In other countries, like New Zealand and India, the test results of students are posted for public review. From eighth grade through twelfth grade, the results of the Department (or Office) of Education's NAEP test for math and science should be posted for the public to see.

I know, I know. The American Civil Liberties Union and other defenders of privacy will resist this radical idea. But in my way of thinking, a little bit of peer pressure could produce big results.

So let's recap. Who's in on the accountability for a successful implementation of the national education strategy? Governors, business leaders, parents, and students. Who's out? The president, Congress, and teacher unions. In my mind, it is that simple.

#3: Innovation Must Be "US"

Yuasa's Phenomenon, which claimed that since 1540 the world's scientific center has shifted from one country to another every 80 to 100 years, was possibly the most interesting thing I learned in writing this book.

Mitsutomo Yuasa, who authored this theory in 1962, placed the mantle of "world scientific leader" on the United States starting in 1920. If Yuasa's Phenomenon is in play again, a shift away from America as the world scientific leader could happen soon, at least by 2030.

Some believe it is already happening. Tom Friedman, the best-selling author and *New York Times* columnist, has written widely on the topic of America's competitiveness. On October 7, 2010, he wrote a column in the *Times* entitled "America's Decline," in which he told the readers that he had done a search on those two words and gotten 42,500 hits.

I was intrigued. Two years later, on October 23, 2012, I, too, did a search on "America's decline" and got 197,000 hits! It seems the decline of America is a topic increasing in popularity.

Friedman's and my searches were based on the algorithms of a search engine. The most authoritative and respected source of global competitiveness, and America's decline or resurgence, is the World Economic Forum's annual *Global Competitiveness Report*, which ranks the competitiveness of 140 nations by scores of data points.

The *Global Competitiveness Report, 2012–2013* placed the United States seventh among all nations measured. Some might think that 7 out of 140 isn't so bad. It is, however, when six years ago the United States was ranked first in the world in the World Economic Forum's report.

Author Jim Collins, in his book *How the Mighty Fall*, offers five stages of organizational growth and decline, which I believe can also be applied to countries. Here are Collins's stages, with my parenthetical observations of when the United States was, is, or will be in each stage:

1. Hubris born of success (post–World War II)
2. Undisciplined pursuit of more (late 1980s to 2000)
3. Denial of risk and peril (2000 to present)
4. Grasping for salvation (2030)
5. Capitulation to irrelevance or death (2050)

On October 18, 2011, the *Economist* published the article "America Must Manage Its Decline," in which writer Gideon Rachman said, "If America were able to openly acknowledge that its global power is in decline, it would be much easier to have a national debate about what to do about it. Denial is not a strategy."

It seems the cards are stacked against America. But in fact they may not be—thanks to Yuasa's Phenomenon.

A core component of Yuasa's Phenomenon is that while one country may be perceived as the world's overall scientific leader, another country could hold the title of "world's best" in a specific category.

The category that America must dominate to remain relevant in the future is *innovation*. America has a long, rich tradition of innovators like Henry Ford, Thomas Edison, the Wright brothers, and Steve Jobs. Our nation is just good at innovation. Although innovation can happen anywhere and be done by anyone, most would agree that the process of innovation thrives in an environment of smart scientists, technologists, engineers, and mathematicians.

Currently, 87 percent of the world's STEM undergraduate degrees are conferred each year to students in China, India, and Brazil. There is no way America can ever match those numbers, even if *every adult* in America earned a science, technology, engineering, or math degree!

But there is an opportunity for the United States to be the world's most innovative country. Even though China graduates hundreds of thousands more students with STEM degrees each year than the American university system does, McKinsey & Company (2005) reports that less than 10 percent of China's STEM graduates are "suitable for work in multinational firms."

To create a nation of innovators, the American education system must morph from a system based on mediocrity to a meritocracy. Children learn at different rates, and our national education system needs to do a much better job identifying student superstars at an early age and proactively move them quickly through the education system after grade five.

Your business does not have a human resources policy in which promotions are doled out solely by age. Neither should our national education system.

As mentioned throughout the book, our K–12 education system must be the best in the world at teaching the five Cs of critical thinking, collaboration, communication, creativity, and confidence. These are the skills that multinational businesses consider suitable for employment in their firms. Over the next two decades, if China or India does a better job than American schools at teaching these important skills, it will make it very difficult for America to become the world's innovation leader.

Future U.S. government funding trends in basic research—unstructured research driven by scientific curiosity and discovery—will also be a key determinant in the ability of the United States to be the world's future innovation leader. Private investment (and thus future jobs) often follows public investment trends. To regain its top perch as the world's leading innovative country, the United States must increase its funding commitment to basic research and make sure those institutions receiving the funds use them for research and development, not overhead. It's that simple.

My favorite innovation benchmark is the annual country-by-country report on triadic patents—patents that are simultaneously approved in the United States, the European Union, and Japan—found easily on the Organization for Economic Cooperation and Development web site, www.oecd.org/sti/oecdsciencetechnologyandindustryoutlook.htm. The United States does a mediocre job with triadic patents, but China is among the worst countries in the world. This gives the United States a strong opportunity to once again become the world's most innovative country.

Alan Kay, an American computer scientist, has summed up well America's innovation opportunity and challenge during a talk at the Palo Alto Research Center in 1971: "The best way to predict the future is to invent it."

America's future depends on our nation's ability to innovate and invent—a goal that requires a population with strong STEM skills.

4. Hire the Best Teachers

As millions of baby boomer teachers opt for retirement in the coming decade, America has a once-in-a-generation opportunity to replenish its teacher workforce.

In Part Two, you read about a McKinsey & Company survey that focused on why relatively small countries like Finland, South Korea, and Singapore continue to be among the best-performing countries in international math and science assessment tests. Before this study, the common wisdom was that Finland, South Korea, and Singapore were small, homogeneous societies (compared with the large and diverse nature of the American population) and that it is easier to instruct students from a similar cultural background.

Wrong! What McKinsey & Company discovered was that these countries had proactive, nationally based teacher hiring policies that recruited new teachers only from the top 30 percent of the country's college graduates (and the top 5 percent for elementary school teachers in South Korea). Moreover, these countries pay teachers well (a $55,000 starting salary) and then offer them a wide array of professional development programs.

It is the polar opposite in the United States, where only 30 percent of new teachers are among the top college graduates. Moreover, the American system offers back-ended, union-crafted compensation packages with a low starting salary ($35,000) but a cushy retirement plan.

McKinsey & Company researchers wanted to know why America's best and brightest college graduates shun careers as teachers. The answer, in one word, is money. The study found that a minimum starting salary of $65,000 was the tipping point to get the top U.S. college graduates to consider careers in teaching.

Where will America get the money to nearly double the average starting salary to hire bright teachers? Read on!

5. Create 50 State Education Trust Funds

Every time you buy gasoline you pay 20 cents or more into a national fund known as the Highway Trust Fund. The Highway Trust Fund, created in the 1950s as President Eisenhower was beginning construction of our nation's interstate highway system, is a usage tax intended to pay for road maintenance and construction. It's simple and straightforward: You drive, you pay (unless you are driving an electric car). If you don't drive, you don't pay.

When I read the McKinsey & Company report's claim that $65,000 was the tipping point to entice America's best college graduates to teach, I came up with an idea: Create 50 state education trust funds to pay for the extra money needed to move the starting salary to $65,000 for any new teacher in America, whether he or she teaches science, math, English, reading, or history!

Here's how I would do it. The two groups with the most vested interest in improving math and science education in our country are businesses, which often complain that our schools aren't doing a good enough job, and parents, who want a successful future and good-paying jobs for their children.

Because I believe that most businesses and households already pay their fair share of taxes, I would recommend that these 50 state education trust funds be created through this formula:

Business: $10 annual tax per worker employed in a particular state

Parents: $10 per child ages 6 through 17

I estimate that this formula would generate $4.3 billion each year and pay for hiring 143,000 new teachers—*with one stipulation*. State departments of education could offer $65,000 starting-salary packages to college seniors who wish to become teachers and whose grade point average is within the top 20 percent of their college graduating class.

I know that grade point average is a very subjective measurement that differs widely from one college to another. But we have to start somewhere in convincing our nation's smartest college graduates to opt for careers as math and science teachers.

Some will criticize this unorthodox idea. "Wait a minute," they will say, "my brother has been a public school teacher for 20 years and only makes $55,000. This idea is not fair."

My retort would be that yes, a teacher working 20 years is probably making in the mid-$50,000 range. But that same teacher also has 20 years of traditional pension plans built up. Teachers accepting the $65,000 starting salary would have no pension fund, only a traditional 401(k) account.

I cannot overstate this opportunity for America. With millions of baby boomer teachers leaving the profession in the coming decade, we absolutely must fill these open positions with the best and brightest math and science minds coming out of college. If we don't, there is no possible way our country will reach the 2033 goal of the national education strategy.

Lee Iacocca, the well-known automobile executive, once said, "In a completely rational society, the best of us would aspire to be teachers and the rest of us would have to settle for something less."

We must do all we can to find "the best of us" and place them in classrooms across America. And as you read the next recommendation, consider whether that might mean you!

6. America Needs More Second-Career Math and Science Teachers

A significant challenge facing math and science education in the United States is the high percentage of math and science teachers, particularly in middle school, who teach math and science out of field, which means that they have no undergraduate or graduate STEM degree.

The IBM Corporation's Transition to Teaching program (see Chapter 14) is an excellent model that any business can emulate to help employees, usually with 20 or more years of service to the company and possessing a STEM degree, to transition into a second career as a math or science teacher.

Teach for America, the high-profile and well-respected nonprofit group that places college graduates in teaching positions in urban and rural areas across the United States (see Chapter 13), is also actively recruiting second-career candidates. It is remarkable that 15 percent of Teach for America's new teacher hires are second-career teachers, a statistic that has grown in the past decade.

7. Free the Best Teachers from Zip Code Jail

American education is a collage of 14,500 school districts. The best teachers are often found in the more affluent towns because these communities pay more money.

Imagine a day in the life of the best middle school math teacher in America. This superstar probably teaches six classes of 25 students a day, thus working directly with only 150 students each school year. What a dramatic waste of talent. I call this "zip code jail."

The 2000 report *Before It's Too Late* (see Chapter 11) recommended the creation of 15 teaching academies, spread across the country, where great teachers could meet and share ideas about teaching methods with proven results. We also learned about the Khan Academy, a highly popular online teaching model (see Chapter 14). Why not combine them?

To free the best teachers in America from zip code jail, I propose melding the academy described by *Before It's Too Late* with the Khan Academy to create 15 math and science teaching academies. Every six months, 100 outstanding middle school math and science teachers would enroll and create online math and science lessons that would help millions of middle school teachers and students learn and master math and science outside normal school hours.

The National Science and Math Academy, as I would call it, would recruit math and science teachers from within a 40-mile radius of each academy so the attending teachers could commute to and from the job. The teachers would continue to receive their normal pay while attending the academy.

The National Math and Science Academy would allow tens of millions of middle school students in America, no matter what zip code they live in, to have a direct relationship with the best and brightest math and science teachers in our country, hence freeing these teachers from zip code jail.

Why just middle school? My review of decades' worth of NAEP and TIMSS data points to one ominous trend: American student performance in math and science slides precipitously from fourth grade to eighth grade. That worries me, because it should be just the opposite. In fourth grade students have one teacher who teaches everything, but in eighth grade they are enrolled in classes, like science and math, that have subject-specific teachers. The problem, as recounted throughout this book, is that a significant portion of middle school math and science teachers have no academic degree in science, technology, engineering, or math.

By focusing the initial efforts of the academy on middle school math and science, I hope the idea of superstar math and science teachers could help out-of-field middle school math and science teachers do a better job in the critical career-forming middle school years.

8. Create IT Job Ambassadors

A significant challenge to STEM education in America is the misperception that young students (particularly girls) harbor about science, math, and IT careers. We also noted that this misperception is often enhanced by guidance counselors.

I want every IT professional reading this book to meet with your local middle school or high school guidance counselor. Borrowing from the Massachusetts-based DIGITS program (see Chapter 14), become IT job ambassadors.

At the meeting, explain what an IT worker does for a living and cite Bureau of Labor statistics that report job employment trends in the IT industry for the coming decade. Offer your willingness to mentor a student interested in learning more about the IT profession. Explain to the guidance counselor about the great opportunities for women in STEM careers.

There are 99,000 public schools and 5 million IT workers in America. That is over 50 tech workers for each school. I like those coverage odds.

9. Grades 13 and 14 Are Critically Important

As noted earlier, in 2012, 30 percent of adult Americans had earned a college degree, the highest rate ever in our country's history.

Yet equally daunting are these stats from the U.S. Census Bureau: 39 percent of American adults have only a high school degree, 8 percent have less than a high school diploma, and an amazing 5 percent never made it past eighth grade.

Although the national high school dropout rate of 8 percent is commendable, it should still be a cause for concern, particularly when combined with the number of Americans who only make it to eighth grade. Georgetown University's Public Policy Institute has noted that by 2020, 66 percent of all jobs in America will require an individual to have a postsecondary degree.

When I was growing up, postsecondary schools that weren't colleges were called vocational schools, and they had a stigma as the destination for those with mechanical, not intellectual, skills. That's no longer the case, as shown by groups like Skills USA (Chapter 13), CompTIA (Preface), and the remarkably innovative P-TECH (Chapter 15), which starts in ninth grade and continues to fourteenth grade, when graduates earn an associate degree in computer science and engineering, a program that President Obama highlighted in his 2013 State of the Union address.

The key issue in this recommendation is that 12 grades are no longer enough. Tens of millions of future jobs in our country will demand grade 13 and grade 14 STEM skills.

10. The Five Cs Must Join the Three Rs

Throughout this book I have addressed the importance of the five Cs: critical thinking, collaboration, communication, creativity, and confidence. These should be taught as basics in our nation's curriculum, starting in kindergarten, along with reading, writing, and arithmetic.

I have seen the outstanding work that groups like the Junior FIRST Lego League (Chapter 14), Engineering Is Elementary (Chapter 14), and the Tech Corps (Chapter 13) have done taking basic STEM principles to grades one through three.

I believe strongly that the American education system should consider beginning to teach math and basic science as individual subjects starting in third grade.

SAS Curriculum Pathways, the Cisco Networking Academy, and Intel Teach (all in Chapter 13) are excellent examples of corporate initiatives to foster teaching the five Cs. So, too, is the Khan Academy (Chapter 14).

In the twentieth century, reading, writing, and arithmetic were enough for a workforce wedded to manufacturing. But they're not enough in 2013. Workers need to think critically, collaborate with other workers around the corner and around the world, know how to clearly communicate ideas, and be creative and confident.

How do you incorporate the teaching of these skills into a formal curriculum on a cloud-based services model? It is not easy. But how we do it, or don't, just might determine the future economic success of our country.

Closing Time

I have hardly addressed everything that an IT executive needs to know about the state of STEM education in America. But this book should serve as a guide to the major events, studies, research reports, and test scores of the past 50 years that explain how America arrived at its current situation, and you should now have a sense of the challenges the United States faces and how you, as an IT executive, can and must get involved to help craft the solution.

To secure the American "empire of the mind" (to use Churchill's phrase) in the twenty-first century, we must win those battles of the mind. But by several key indicators, we are not doing well. In quantitative international science and math assessment tests, American students perform mediocre at best. In domestic science and math tests, at least 70 percent of American students are rated below proficient in math or science.

The World Economic Forum's *Global Competitiveness Report, 2012–2103* reveals the United States has fallen from first place in 2007 to seventh place in the current study. According to that report, America's science and math teachers rank forty-seventh in the world in quality.

The American public knows that our public education system is failing. Yet parents, teachers, administrators, union leaders, and politicians seem to act like the proverbial frog in the pot of water that remains content even as the water temperature slowly rises to a boil.

One of the most provocative reports I read while researching this book was *America's Choice: High Skills or Low Wages!* (Chapter 10). Twenty-three years later, the title of that report bluntly addresses a painful question for the United States.

America, we still have a choice—although time is running out, if you put any credence in Yuasa's Phenomenon. Will we systemically restructure our education system to teach the critical math and science skills that will ensure our country's future economic, employment, and national security? Or will we remain content with the current system mired in the teaching methods of the nineteenth and twentieth centuries and dominated by teacher unions? It is *our* choice—and right now our country is lucky to still *have* a choice.

As I was finishing this book, I had the opportunity to meet with former President Bill Clinton at a technology conference. Because I was one of the keynote speakers at the conference, I was invited to a special photo session with him. I approached President Clinton, and we shook hands. I thanked him for helping me to launch the nonprofit Tech Corps in 1995, and as I looked into his eyes, I could tell he had probably forgotten about the October 1995 Oval Office event. As I started to walk away so that the next person could meet him, I felt a gentle tug on my left elbow. President Clinton drew me closer to him, leaned over, and whispered, "We did good, didn't we?"

I was ecstatic that he remembered, and that story serves well the closing thought of *The U.S. Technology Skills Gap.* Every IT executive who reads this book has the same opportunity to do good.

One of my favorite Chinese proverbs is "I hear and I forget. I read and I remember. I do and I understand."

For nearly 60 years, millions of Americans have *heard* from thousands of sources that the American education system, especially how it teaches math and science, is broken. But most have *forgotten* those reports.

It is my hope that your *reading* of this book will help you to *remember* the seriousness of this national challenge. If we do not meet this challenge, it will negatively affect our country's future economy and national security.

But the highest compliment you could pay me for the six years of research that went into *The U.S. Technology Skills Gap* is for the content of the book to inspire you to *do* something about it. For only then will you truly *understand* the gravity of the issue.

ABOUT THE AUTHOR

Gary J. Beach is Publisher Emeritus for International Data Group's *CIO* magazine, an online and print publication written for chief information officers. During his career at *CIO* he has served as the magazine's primary spokesman, appearing frequently in media organizations such as *USA Today*, the *New York Times*, National Public Radio's "All Things Considered" program, where he contributed technology commentaries for nearly five years, and CNBC's "Squawk Box" and "Squawk on the Street" financial news programs, where he reported technology investment forecasts. He and his wife, Catherine, and their cat, Zorro, live in the Boston area.

ABOUT THE WEBSITE

This book includes a companion website. You can access it via this link: www
.wiley.com/go/USTechSkillsGap.

The website contains the bibliography for the book as well as a Further
Reading section, which contains a list of valuable resources selected by the
author. The list is grouped into six areas for easy referral.

The password to enter the site is Beach.

INDEX

Page numbers in *italics* refer to figures or tables.